1982

The Organization & Administration of Emergency Medical Care

George Sternbach, M.D.
Assistant Medical Director of Emergency
    Medical Services
Assistant Professor, Department of
    Surgery
Stanford University Medical Center,
    Stanford, CA.

a TECHNOMIC® publication

TECHNOMIC Publishing Co., Inc.
265 Post Road West, Westport, CT. 06880

The Organization & Administration of
Emergency Medical Care

©TECHNOMIC Publishing Co., Inc. 1978
265 Post Road West, Westport, CT 06880

a TECHNOMIC publication

Printed in U.S.A.
Library of Congress Card No. 78-74611
I.S.B.N. 087762-270-1

FOREWORD

This volume is an edited transcription of the discussions presented at "The Organization & Administration of Emergency Medical Care", a postgraduate medical education seminar held in Chicago on May 4-6, 1977. Although no effort has been made to provide a verbatim transcription of the presentations, a true account of the proceedings has been undertaken. The work on this volume was performed while the editor was an Assistant Professor in the Division of Emergency Medicine at the University of Chicago Hospitals and Clinics. We appreciate the help of the contributors in aiding with the preparation of the text. We also gratefully acknowledge the work of Ms. Laurelle Battaglin and Donna Kanelopoulos for secretarial support and Mr. Walter Knish for editing the manuscript.

**Dedicated
To My
Parents**

CONTRIBUTORS

Frank J. Baker, II, M.D.
Assistant Professor & Director
Division of Emergency Medicine
University of Chicago Hospitals and Clinics
Chicago, Illinois

Thomas P. Cooper, M.D.
Medical Director
Physicians Placement Group, Inc.
St. Louis, Mo.

Lawrence R. Drury, M.D.
Staff Physician
Emergency Medical Services
Denver General Hospital
Denver, Co.

Peter Rosen, M.D.
Director, Emergency Department
Denver General Hospital
Denver, Co.

Marshall B. Segal, M.D.
Associate Professor
Division of Emergency Medicine
University of Chicago Hospitals and Clinics
Chicago, Il.

Carole P. Swinehart, B.S.N.
Region III, M.I.C.U. Co-ordinator
University of Chicago Hospitals and Clinics
Chicago, Il.

Michael C. Tomlanovich, M.D.
Senior Staff Physician
Division of Emergency Medicine
Henry Ford Hospital
Detroit, Mi.

TABLE OF CONTENTS

CHAPTER 1

REGIONAL DISASTER PLANNING

Frank J. Baker II, M.D.

The design of an adequate disaster plan is dependent on the knowledge of physicians, administrators, and government officials that disasters are not rare and that the formulation and practice of a rational and smoothly functioning plan will save lives during any disaster.

A brief survey of Chicago history demonstrates adequately that disasters are not rare occurrences. The Chicago Fire occurred in 1871, and resulted in 250 deaths, countless injuries and more than 17,000 destroyed buildings. In 1903, the Iroquois Theatre burned down and 602 people died in that fire. In 1915, a weekend cruise ship capsized in the Chicago River with several thousand people aboard. A local corporation had hired the boat for a company picnic, and the excited passengers all went to the side of the ship nearest the pier to wave good-bye to their relatives; the ship rolled over and capsized, and 812 people drowned. Several hundred more lives were lost during the subsequent years in small scale disasters, such as the Chicago Stockyard's fire and the La Salle Hotel fire. In 1958, Our Lady of Angels school burned with

1

the resultant deaths of 58 children.[1]

In 1970, the Center for Health Administration Studies of the University of Chicago published the Gibson Report, which outlined as one of its recommendations the formation of a city-wide commission to address the problem of emergency medical services for metropolitan Chicago.[2] One of the tasks of this commission was to formulate a functional disaster plan to include all participating agencies within a specific geographical area. The Emergency Medical Services Commission of Metropolitan Chicago was formed, and the Disaster Preparedness Committee was constituted. Work began none too soon on a disaster plan for the city since, early in 1972, two Illinois Central trains collided, which resulted in the deaths of 44 people and 400 injuries. Since that disaster, there have been seven separate disasters which have resulted in over 1,100 victims, and over 100 fatalities in the Chicago Metropolitan area.

From these statistics, it is obvious that there have been a large number of incidents within the Chicago metropolitan area in recent years which can qualify as disasters, both in terms of loss of life and loss of property. The experience of this city has been somewhat unique in that Chicago has had 8 disasters in 5 years, which has allowed the testing and subsequent affirmation or re-design of various aspects of the city's disaster plan. In general, it can be said that there has been an increased incidence of technological disasters within recent times. As more advanced forms of transportation and housing become available, the danger of malfunction increases. There obviously were no airplane crashes prior to the age of flight, and no motor bus accidents prior to the invention of the internal combustion engine. As society becomes more sophisticated and more urbanized, and our technology becomes more complex, the chances of major disruption in our technology increases as well. As our society moves to mass transportation, mass housing, and mass gatherings, the incidence of threats to the lives of large numbers of people also increases. There is no reason to believe that the incidence of natural disasters will decrease over the long run, and thus the net incidence of disasters with the production of large numbers of casualties should increase.

Since disasters are frequent and contingency plans help to deal with the disaster, it is easy to understand the rationale of a commitment to design a functional disaster plan. There are two basic steps in accomplishing this: the first, is to identify and assemble all of the agencies involved in a disaster. The second is to sit down and actually formulate the disaster plan based upon the expertise of the individuals and agencies involved.[3]

In accomplishing the above, a number of tasks must be completed. An overseer agency that will take responsibility for putting the plan together must be identified. This may well be a different agency than the one which has the overall responsibility for running the disaster response. In fact, in the Chicago Metropolitan Plan, the overseer agency is the Emergency Medical Services Commission of Metropolitan Chicago and Cook County, while the agency responsible for the actual management of the disaster is the Chicago Fire Department. Once the overseer agency has been identified, a disaster committee must be formed with constituent members all of the agencies that would actually participate in the disaster.

It is critically important to involve hospital-based physicians in the formulation of the disaster plan. Medical disaster teams will most often be comprised of hospital-based physicians because of their ready availability. Input by the physicians who will deal acutely with the patients during an actual disaster will help in designing a realistic plan, since these individuals will be most knowledgeable of the capabilities, staffing, supplies and procedures of the medical agencies involved. All too often, regional disaster plans are formulated by administrators without medical input at all, or with input from physicians who have no active role in the actual management of the medical aspects of the disaster.[4] Physicians also need to be involved in deciding the line of command and authority at the scene since patient welfare must come before other aspects of the disaster management, such as traffic control and site security.

After identification of all of the participating agencies, their functional role in the actual management of the disaster must be determined. Some organizations, such as the Illinois Department of Public Health, and the Chicago Hospital Council have a role mainly in the development of the disaster plan, but have little to do with the actual management of the disaster. Key agencies will include the police department, fire department, city and state departments of health, civil defense, the Red Cross, participating hospitals, utility companies, the Atomic Energy Commission, regional blood programs, and finally, the military.

After eliciting the aid and cooperation of the agencies involved, identify the role of each one of the participating agencies in the actual management of the disaster. Identification of roles is key to the smooth operation of a rationally constituted disaster

4

plan. In the City of Chicago, the Police Department has primary concern with crowd and traffic control, while the Fire Department is in charge of overall disaster management. The City Department of Health, in conjunction with the Chicago Fire Department, is responsible for the medical aspects of the disaster, while the Illinois Department of Public Health will usually not be involved, but can provide certain manpower and equipment resources to the operational field. For instance, the state can provide state police manpower vehicles and helicopters. Civil Defense, which is an operational arm of the Chicago Fire Department, is particularly suited to provide equipment and supplies to the disaster site. The Red Cross is a very useful organization that has provided critical back-up communication systems during disasters, and has been key in keeping tabs on the location of victims after they are removed from the disaster site. In the Chicago plan, the American Red Cross uses the public information network to broadcast their local telephone numbers where relatives and friends can call regarding information about victims. The private ambulance sector can be of great help, and usually will provide service during a disaster at no charge to the patient. Their ambulance association should be considered as a participating agency for the purposes of planning. They have been most cooperative during our previous disasters, and can muster men and equipment in a short time. To give you some idea of the adequacy of their response, they have routinely been able to send 15—25 ambulances to the disaster site on 3—5 minutes' notice.

Because of the necessity for cooperation from the utility companies, their participation in the formation of the plan is critical. The Atomic Energy

Commission will have to be involved in all cases of nuclear accidents; one should not ignore the importance of their input to the disaster planning process on the premise that nuclear disasters are rare. The cases in which the AEC is most likely to be consulted are those where there is a limited disaster in which a radioactive isotope may be involved. The military also needs to be consulted for advance planning since it will then be assured that, in the event of a major disaster, a local plan will not be superseded by a military plan, which may not have been recently tested and which provides for no involvement or minimal involvement of local officials and organizations. It is quite clear that in a major disaster in which martial law is declared, the military will play a large role in the actual management of the disaster site.

The identification of a geographical area of primary responsibility will permit the identification of all of the governmental agencies involved in managing a disaster within that area. It will also permit identification of the different types of potential disasters. In the City of Chicago, we have put together a map of the city identifying plants using radioisotopes, explosives, high energy compounds, and dangerous chemicals. Using this map, we can identify potential major industrial accidents, and predict possible propagation of these diasasters.

Because the agency with overall responsibility for running the disaster may lack medical sophistication, it is the responsibility of the participating hospitals to provide the necessary medical expertise to insure the adequacy of the plan from the standpoint of medical care for the victims.

Once all of the participating agencies have been defined and their responsibilities outlined, each should establish a set of standard operating procedures to be used in the event of a disaster. This "cookbook" needs to be clear, concise, and sufficiently detailed to allow for the activation of the disaster protocol in case an individual unfamiliar with the disaster plan must activate it.

Concepts of modern disaster management call for rapid access to the public protective agencies, and rapid response by these agencies to the disaster site. In Chicago, this is supplied by the institution of a 911 system, which is staffed by the Chicago Police Department. The communication system is designed such that there is rapid transfer of pertinent information to the Chicago Fire Department telecommunication center (TCCC), which has the responsibility of maintaining all communications during the disaster.

The disaster plan must be easily activated and must be adaptable to many situations. Activation is by any battalion fire chief. Variable response of the plan has been assured by labeling the disaster by the numbers of victims involved. Plan I is for 5—15 victims. Plan II is for 15—30 victims and Plan III is for 30 or more victims. Medical teams are sent at the request of the battalion chief or higher supervisory personnel or at the discretion of the ranking physician at the hospital command post on the basis of the number and types of casualties involved.

The communications aspect of disaster management is a major task, and the responsible agency (in the case of Chicago, the Fire Department), must maintain communications with its numerous personnel and equipment, while simultaneously coordinating communications with the Police Department, civil defense, military, utility companies, Red Cross, private ambulance sector and participating hospitals.[5, 6, 7] The reality of the disaster situation is that no single agency, including the Fire Department, can provide the communications necessary for a disaster on a routine, 24-hour per day, 365-days per year basis. It is difficult to justify the expense of the extra manpower and equipment to be available on a constant standby basis.

The above communications overload has been demonstrated on numerous occasions. As a result of the inadequacies of the communications between the participating hospitals and disaster site, the hospitals have assumed the responsibility for maintaining this communication link. Previously, the Chicago Fire Department contacted all of the participating hospitals through regular telephone service in order to notify them of the occurrence of the disaster and to obtain information regarding their ability to receive patients. This information was then relayed to the disaster scene, in particular to the ambulance dispatch area, so that rational patient distribution could be made. However, practice has shown that the Fire Department was so busy doing their primary job of management of the disaster from a fire and public safety point of view, that only late in the disaster were they able to free up the communications manpower and equipment to permit the relaying of information from the hospitals back to the disaster site. The revised plan calls for the Chicago Fire Department TCCC to call only the designated command post hospital for that geographical area and to notify no other medical agency.

Because it was necessary for the command post hospital to have communication links with its surrounding participating hospitals, it was most logical to identify the Mobile Intensive Care resource hospital as the permanent command post hospital for each of the three geographical areas comprising the City of Chicago. This process provides that the dedi-

8

cated telephone lines from the Mobile Intensive Care resource hospital to all of the hospitals in its area which receive patients through the Mobile Intensive Care network, will be used as rapid emergency communications systems to fulfill the medical communicaitons necessary during a disaster. Back-up communications will be provided by the MERCY radio system. The ranking Emergency Medicine physician at the command post hospital has the responsibility for choosing which hospitals will send disaster teams, which hospitals will receive patients, which hospitals will remain on stand-by alert, and for notifying the Chicago Fire Department of these decisions. Just as important, the command post hospital collates information on the Emergency Department status and in-patient bed status of the hospitals which are to receive patients so that TCCC may relay this directly to the ambulance dispatch area. This then permits the rational distribution of patients by ambulance dispatch to the participating hospitals.

While this information is being collated, the various agencies that have been notified proceed to the scene. This, of course, includes the medical disaster teams from the selected participating hospitals. These disaster teams consist of a minimum of one physician, one nurse and one technician. These are to be transported to the scene by their own security cars, since police and fire vehicles may be unavailable. At the disaster site, it is the responsibility of the Police Department to maintain traffic and crowd control, while the Fire Department begins management of the actual disaster site. Paramedics and medical disaster teams respond to the disaster; the first problem is identification of these individuals so that they can pass through the police lines. While this is not a problem with the paramedics, it can be a problem for the hospital based disaster teams. Our solution for this was to designate a green-colored hard hat as the recognized medical identification which allows medical personnel access to the disaster site. The reason for choosing the color green was that it is easy to see, and is an unpopular color for the hard hats used by other personnel. The hard hat was chosen as a result of the need for some protective gear for medical personnel while at the disaster site.

Once on the site, medical personnel must be able to easily identify the location of the medical command post. This problem was solved by identifying a special vehicle, the medical triage unit (#472) as the post to which the medical personnel should report. The command post will be staffed by Chicago Fire Department personnel under the command of the medical director of Emergency Medical Services of the Chicago Fire Department.

Once they have reported to vehicle #472, they are to proceed to illustrate the principles of modern disaster management, which call for rapid field triage followed by field stabilization, prior to rapid distribution through a geographically identified ambulance dispatch area.[8, 9, 10, 11] The necessity for setting up and transporting patients from the field to specific field hospitals prior to removal from the disaster site will depend upon the extent of the disaster in terms of the number of casualties and types, and the adequacy of the personnel and facilities responding. It must be noted that paramedics and disaster teams responding to the disaster report to the medical command post to do rapid field triage under the supervision of the ranking physician on site. This is the medical director for Emergency Medical Services of the Chicago Fire Department; in practice, such an individual has not always appeared on the scene. The first physician responding as part of a hospital based medical disaster team should assume responsibility for operation on the on-site medical command post, and for supervising the rapid field triage and subsequent field stabilization, as well as the assembly of field hospitals and the geographical ambulance dispatch site.

Medical and paramedical personnel then begin rapid field triage aimed at categorizing patients within the first few minutes of the disaster. In most circumstances the paramedics will be the first medical personnel on the scene, and in disasters with small numbers of casualties, the medical aspects of the disaster may be secured by the time the hospital based medical disaster teams arrive. For these reasons, the paramedics have been trained in rapid field triage. This, combined with their knowledge of field stabilization, allows them to proceed with field triage, field stabilization and rational ambulance dispatch immediately when they arrive on the disaster site, despite the fact that there may be no physicians in attendance. At this point, only the very basics of first aid, such as tamponade of arterial bleeding and respiratory assistance to victims will be given. Only through a limitation of the medical treatment at this stage can the seriousness of injury to all the victims involved be assessed at an early time. This rapid categorization of patients then permits the use of manpower and resources to the maximum benefit of the greatest number of victims. The triage system and triage tag adopted by the City of Chicago is a four-category system designed by the Journal of Civil Defense from Stark, Florida. It consists of categories 0 (black), 1 (red), 2 (yellow), and 3 (green). Patients labeled #3 (green) are the ambulatory wounded whose injuries are such that they do not even need transportation on an emergency basis. Patients labeled #2 (yellow) are those who are seriously injured and require stabilization prior to transportation to the hospital, but whose injuries are such that their immediate survival is not dependent on immediate care. Patients in category #1 (red) are those who are critically injured, whose lives are immediately threatened as expectant dead may be placed in either the deceas- (black) are deceased. Patients who might be labeled the expectant dead may be placed in either the deceased category or the critically injured category, depending upon the personal feelings of the physician or the paramedic personnel involved, as well as the extent of the disaster and the type of casualties. A 90% burn in the Coconut Grove fire may well have been placed in the black triage category, whereas in a rather limited disaster involving only 10 or 15 victims, where no one else is seriously injured, this patient would be placed in the red triage category.

The triage tag used must be simple and of rugged construction, and labeled such that personnel unfamiliar with the tag can, within a few minutes, learn to use it adequately. The tag designed by the Journal of Civil Defense meets these critieria.

Once rapid color-coded triage has been done, the available personnel must then immediately attend to the needs of those patients categorized as critical. This will require the use of drugs and equipment which will ordinarily be available on the Mobile Intensive Care Units. These drugs and equipment will also be provided in the disaster trunks accompanying the hospital based medical disaster teams. Subsequent re-supply, if depletion occurs, will be provided by the hospitals at the request of the ranking physician at the disaster site through the TCCC command post hospital communications network.

After field stabilization, patients will be moved either to a field hospital or to a geographical ambulance dispatch site, depending upon the number and types of victims involved. If field hospitals are set up, this will be done at the discretion of the ranking physician and staffed by a combination of the medical disaster teams and paramedics available. There will be a separate field hospital corresponding to each triage category with the category 1 field hospital closest to the ambulance dispatch area.

Ambulance dispatch will be staffed by members of the Chicago Fire Department's Bureau of Emergency Medical Services, preferably with the help of a physician. Through walkie-talkie communications, they will receive constant information regarding Emergency Departments and hospital bed availability, in order to permit them to make rational decisions about the destination of victims transported from the disaster site.

At the conclusion of the disaster, a wind-down of participating agencies, men, and equipment must be accomplished. Key in this is the identification of the agency that has the responsibility for calling off the disaster mechanism and for notifying all of the agencies involved.

Once formulated, the plan should be practiced on at least an annual basis. The Joint Commission on Accreditation of Hospitals now requires four hospital disaster drills per year. Our area-wide Emergency Medical Services Committees are now moving towards having two regional disaster drills a year. The Joint Commission will accept these as fulfilling part of the requirement of each participating hospital to test its own internal disaster plan. Clearly, the disaster plan must be tested and practiced regularly if it is expected to operate during a real disaster.[12]

The Metropolitan disaster plan has been regularly tested with pre-scheduled drills, and at least two unscheduled drills will occur this year. Next year, we plan to move to unscheduled night disaster drills, in order to more accurately test the function of the plan. Fortunately, many disasters are industrial and transportation-related, and thus occur during the daylight hours when hospitals are at full staff. It is, however, necessary to pinpoint the problems that occur when a disaster happens at 3:00 A.M., when hospital staffing is at a low point.

Disaster planning is an integral function and responsibility of Emergency Medicine. The medical aspects of disaster planning must be developed after specific input from hospital based physicians, and coordinated with plans of operation involving civil agencies, such as the police and fire departments, if the medical care given during an actual disaster is to be optimal. In addition, hospital based personnel, including physicians and nurses, are necessary to determine which hospitals will respond, the level of the response, and to collate and relay information about hospital status to the field. They are also needed on site to provide maximum medical care to the victims. The plan must be tested under a variety of conditions to assess its actual functional capabilities.

REFERENCES

1. Statistics, Chicago Hospital Council Official Disaster Critique Records.
2. Emergency Medical Services in the Chicago Area, Gibson, Center for Health Administration Studies, University of Chicago Press, 1970.
3. Hollis TL and Sapp BW: The Hospital as an Emergency Center, Hospitals, J..A.H.A., May 1, 1972.
4. King EG, The Moorgate Disaster: Lessons for the Internist, Annals of Internal Medicine, 84: 333—334, 1976.

5. Deisher JB, Preplan Your Disaster, GP, December, 1965.

6. Owens JC, Emergency Health Services Require Efficient Communications System, Hospitals, J.A.H.A., June 16, 1969.

7. Allenbaugh GE, Emergency Radios Restore Order to Chaos, Hospitals, J.A.H.A., January 16, 1965.

8. Rutherford WH, Experience in the Accident and Emergency Department of the Royal Victoria Hospital with Patients from Divil Disturbances in Belfast 1964-72, with a Review of Disasters in the United Kingdom 1951-71. Injury 4: 189-199, 1973.

9. Hart RJ, Lee JO, Boyles DJ, et al. The Summerland Disaster, British Medical Journal 1: 256-259, 1975.

10. Holloway RM, Medical Disaster Planning: Urban Areas, New York Journal of Medicine, March 1, 1971.

11. Conrad MB and Klippel AP, Disaster Planning in a Metropolitan Area, Bulletin of the American College of Surgeons, May, 1972.

12. Bohn GA and Richie CG, Learning by Simulation: The Validation of Disaster Simulation: Medical Scheme Planning, Journal of the Kansas Medical Society, November 1970.

CHAPTER 2

EMERGENCY DEPARTMENT DISASTER PLANNING

Peter Rosen, M.D.

Fortunately, I have not personally had to respond to a true major disaster — such as an atomic explosion or a major earthquake. I am typical in this respect. Most of the disasters with which the ordinary emergency physician will have to deal are those in which the total number of victims involved is less than five hundred. Nevertheless, there is a significant disruption of orderly activity when even a large hospital is suddenly flooded with several hundred patients.

An essential part of disaster management is having an existing plan that is well known to key individuals of an Emergency Department staff. An important component of the disaster plan is communications. There should be some pre-established method by which the news of a disaster is relayed to a hospital. Communication is also involved in the distribution of patients from a site. The best component of a hospital disaster plan is the management of the disaster scene. A properly functioning disaster squad can divide accident victims among hospitals so that no one hospital is overloaded. In a city the size of

Chicago, this might mean that no hospital will get sufficient numbers of patients to even justify activating intra-hospital disaster plans. A disaster squad should be headed by a physician, preferably one who has experience in disaster management. He must be prepared to take charge of the triage priorities at the disaster site.

In organizing a response to a disaster, communications are very important *inside* the hospital. Intra-hospital communication includes notification of essential personnel within the hospital that a disaster has occurred, as well as calling additional personnel into the hospital. The latter is essential, especially if the disaster occurs at night or on a weekend, times when staffing is low. There should be a pre-arranged call-up procedure put into operation. This procedure should have been practiced beforehand. You may expect such a call-up to be effective in contacting only about 80% of the needed personnel. Part of the reason is that telephone numbers will be out of date. You need to conduct a call-up drill on a quarterly basis in order to keep both intra- and extra- hospital phone numbers current.

A key person to notify is an administrator who is high enough in the hierarchy to be able to make the institution function under unusual circumstances. He must be acquainted with the location of the needed equipment within the hospital, be aware of disaster procedures and be able to direct non-medical personnel. If there is an administrator on duty in your hospital at all times, this role will fall to him. If this is not the case, an administrative person who can come to the hospital rapidly must be kept on call.

Advanced notice regarding the numbers of patients arriving and the severity of their injuries is helpful, but not always available. An arrangement must be established to create space for the influx of patients. There are usually a number of hospital inpatients who can be discharged to make room for disaster victims. A mechanism to facilitate the immediate discharge of such patients must exist. This presents a difficult problem in theory, since patients and families must be notified and beds evacuated on very short notice. In actuality, however, it works out more smoothly than one would expect. We have had to resort to this several times and have not had a single complaint from a patient whom we have summarily discharged. In the Emergency Department itself, there may also be a number of patients at any given time whose medical problems are not so acute that treatment cannot be postponed. Such patients should be informed that a disaster has occurred and that the area is to be devoted to treatment of victims for a number of hours. The patients should be asked to leave and to return at some later time. Again, I have experienced very good cooperation from patients in this regard.

The Emergency Department will be capable of different kinds of responses, depending on its size and staffing. The role of an Emergency Department staffed by a single physician should probably be minimal. It should continue to function as usual and have little or nothing to do with the influx of disaster victims. This is difficult to arrange, because the Emergency Department is the only part of the hospital that is geared towards handling emergencies at all times. Many emergency departments will not be of sufficient size to accommodate large numbers of patients. In our own experience, we have utilized the hospital lobby areas for the temporary situation of non-life threatened patients. Outpatient clinic areas are also useful for the screening and initial treatment of people who are not critically ill.

Surgeons have no difficulty in seeing a role for themselves in disaster management. Our problem in the past has consisted of having too many surgeons in the emergency area wanting to perform initial stabilization functions and not enough willing to remain in the operating room until patients were sent to them. Therefore, we often experienced the usual problems encountered when too many physicians care for a given patient.

During World War II and the Korean War, military internists functioned as emergency physicians: They took care of the initial resuscitation of battlefield casualties while the surgeons remained in the surgical hospitals. This worked well once the internists became adept at battlefield resuscitation. Unfortunately, in civilian practice, many internists do not feel comfortable in handling the initial resuscitation of severely injured patients. You can frequently call upon physicians from the obstetrics and gynecology department for certain useful surgical and emergency skills. For the most part, however, they are not readily available because they must continue to care for patients in the delivery suits.

A major problem during a disaster is that of identification of injured patients. Patients should be labeled at the disaster site in case they become incoherent or unconscious by the time they arrive at the hospital. Patients who have lapsed into unconsciousness during transport and who have not been identified previously may remain anonymous for days. The turmoil that this causes among relatives is considerable. In general, an effort has to be made at the hospital to deal with relatives. An area should be established where relatives can wait and where they can receive news and announcements. There are often many more relatives than patients, so that this produces a great stress on waiting facilities.

A plan must also include provisions for handling the media. There was a time when this group could merely be banned from the hospital, but this is no longer possible. I found that television news teams in Chicago often found out about disasters before our hospital did. An area for press and television people must be considered. It cannot be too far away from the medical working area, because this would not suit them. On the other hand, it cannot be too close to where you are administering treatment because both your and the patient's privacy needs to be protected.

The number of phone calls that come into a medical center following a disaster is astonishing. I can't tell how people get the phone numbers of the Emergency Department, but such calls can completely tie up your telephone lines and make it difficult for necessary communication to proceed. You may even consider a dedicated phone line — whose number is not publicized — which can be used for essential communications, in addition to telephone numbers that can be publicized for relatives and friends to call.

An aspect of disaster management that is often not anticipated is the paper work. The paper work that ordinarily goes on for every patient must somehow be completed for an influx of disaster victims as well. In addition, people involved in a disaster must be fed. The additional staff that is called in will have to be fed so that it may continue to function efficiently. You can imagine the stress on a hospital kitchen upon which a demand for 100 extra meals is suddenly made. In response, you may consider having available an alternate catering service that can be called upon in case of an emergency. This need not be more than a local delicatessen that can deliver 100 sandwiches.

Standards for medical care that are different from those in force during normal times may have to be established during a disaster. The major priority during a disaster is defining critically ill but salvageable patients. Patients who are not seriously injured should not receive a complete medical work-up for unrelated problems at this time. The responsibility of disaster medical care is identification of life or major health threats only. This should be transmitted to patients as well as documented on medical records. If there is a large number of critically ill but potentially salvageable patients to be cared for, a great deal of time and effort cannot be directed towards patients whom you reasonably expect will die. This is an attitude that goes against the traditional ethic of physicians. It must be recognized in advance, however, that this sort of decision-making will be at work, especially in very large disasters.

A word about disaster drills: These are now being required by the Joint Commission on Accredidation of Hospitals. There are some difficulties with disaster drills, however. To really shut down normal hospital activity and make the institution function for a disaster drill is something that your medical staff will never want to do. Surgeons will not want to cancel elective surgery for the day, for example. It is very difficult to get the cooperation that is required to truly test your system. There is a purpose to drills, but I see this more fully realized in testing the ability to respond outside of the hospital than inside it. The accurate testing of a hospital's response to a disaster through a drill is possible in the area of communications only. There is a more useful role for the disaster drill outside the hospital. Our drills have pointed out many weaknesses in our disaster response system.

A plan is also required concerning disasters occurring within the hospital. It must be basic enough to be useful and not so detailed as to be unwieldy. The major component is another hospital that can take over a certain amount of patient care. You need the capacity to evacuate patients from your hospital to another that can supply the function that has been lost due to the disaster. If this is an explosion in an operating room, for example, the backup hospital must take over your operating capabilities. For a hospital — such as one located in a rural area — which does not have a nearby institution than can serve such functions, there must be portions of the hospital designated as alternative function areas. The most difficult problems arise when patient areas are involved — which they often are. In cases of this sort, you must be able to respond to the needs of the patients injured in the disaster as well as other patients who will have to be evacuated in order to continue to receive proper medical attention.

There is no way to write a disaster plan that will take into account every disaster that could befall. Obviously it is wise to consider regional geography to determine the sort of natural disaster that will be most probable. There is no doubt that prior planning is helpful. Los Angeles did a good job coping with its recent earthquake because that city had put a lot of energy into its disaster planning prior to that occurrence. Despite a well-formulated disaster plan, you still need the ability to innovate at the site. A disaster plan should be no more than a frame on which to build the body of disaster response.

CHAPTER 3

RADIOTELEMETRY & MOBILE INTENSIVE CARE

Frank J. Baker II, M.D.

Seven years ago at the Center for Health Administration at the University of Chicago published a report entitled "Emergency Medical Services in the Chicago Area" which severely criticized ambulance services in the city of Chicago. It pointed out the ineffectiveness of the existing ambulance service as well as its inadequacy in terms of number. At that time there were approximately 33 vehicles serving a metropolitan population of nearly 7 million. There were so few vehicles and personnel in Chicago that some ambulance vehicles in the city literally never shut off their engines. This report was the impetus for the organization of mobile intensive care in the city of Chicago.

The patient's entry into the mobile intensive care network is through the 911 system. There is no question that this represents the easiest access for the patient into the emergency medical services system. 911 calls are currently answered by a specially-trained police dispatcher. He decides what type of response is necessary: whether this be fire, police, ambulance or other, based on the caller's answers to a few key questions. The phone number and address of the origin of the call is automatically traced and is available about 10 seconds after the call is placed. In the future the plan is to have paramedics perform this function. In effect, what this will represent will be a triage system aimed at more rational utilization of men and equipment.

The metropolitan Chicago area served by the Chicago M.I.C.U. network now contains about 6 million people. Therefore, the system which was set up in Denver probably would not work here. There, one hospital has radio contact with all ambulances. These ambulances, in turn, transport their patients to that hospital. It would be impossible for a single hospital to handle all the ambulance-transported patients in Chicago. Alternatively, there is the plan used in Los Angeles. Ambulances there are assigned to given hospitals and those hospitals are contacted by their respective ambulances. There are 28 hospitals sharing 8 frequencies with little control over priority of use and a resultant nightmare of airwave confusion. We were unwilling to introduce this system here because there are 90 Class A hospitals in this city. We did not want to create an unmanageable communication network by giving each of these institutions equal communications capabilities. We therefore, established three resource hospitals in various parts of the city whose responsibility it would be to communicate with ambulances within their geographic zone. These hospitals would receive information about ambulance patients and issue orders. Since they alone could not accommodate all the patients transported by such a system, it was decided that patients should be transported to the nearest comprehensive hospital emergency room. This patient-distribution system was already in existance, working well and had behind it the legal authority of a city ordinance. A comprehensive emergency room is one which is staffed by physicians, is open 24-hours a day, and has the back-up of all major medical specialties. There are 55-60 of these in the city.

The city was divided into 3 parts on the basis of population. Region I, the north side, has Illinois Masonic as the Resource Hospital. Region II, the loop and part of the west side, has Northwestern University as the Resource Hospital. Region III is composed of part of the west side and the entire south side, with the University of Chicago Hospitals as the Resource Hospital. There are certain areas such as the center of the city which have a very high density population. Other areas have a relatively smaller population, but are larger geographically. Hospitals in these areas are also more sparsely situated. The problem in such areas would be prolonged ambulance transport time and this had to be taken into account when the dividing lines were drawn. Mobile intensive care vehicles had to be deployed in such a manner that even though the transport time might be prolonged, the ambulance response time was short. The high-density population areas call for a rapid stabilization and transport type of response, while the ambulance crews in the large geographical areas are required to administer more treatment prior to and during transport because of the prolonged transport times.

A problem in the heavy industrial areas is that of trains running through them and impeding the flow of ambulances. Ambulance personnel stationed here would also have to be able to administer therapy for a prolonged period of time in the event that trains prevented their timely return to the hospital.

Each of the hospitals has eight-channel communications capabilities. The Chicago Fire Department Telecommunications Command Center (TCCC) is going to have the responsibility of assigning radio frequency channels to all ambulances using telemetry in the future. Through this mechanism, there will be a single agency which will be allotting channels for medical communications at any given moment. This will prevent the various hospitals from interfering with each other by using the same channel. The system is designed so that the same eight frequencies can be utilized simultaneously on the north and south sides of the city without interference. This is due to a low watt repeater system utilized in the Chicago system. Suburban areas near Chicago which instituted radio communications before the city did utilize high watt systems, and, as a result, we have had their calls interfere with our own.

Resource hospitals are equipped with tape recording capabilities to record all telemetry runs. These tapes are reviewed, kept on record for about a week and then erased. The communications are not taped for medical-legal reasons, but rather to monitor and evaluate paramedic functioning. Electrocardiographic tracings can be recorded and reproduced by this system. Command modules at the resource hospitals are also equipped with switchboard systems, which in the future will be used to hook other hospitals into the telecommunications system in either a monitoring only or a monitoring/broadcasting mode. It is certain that this will become necessary if the three resource hospitals alone are unable to handle a particularly heavy load of ambulance runs.

The cost of such a communications system is considerable. A complete system at each of the resource hospitals costs $250,000. This does not include the cost of TCCC, city wide radio antenna coverage, radio equipment for the paramedics, or their vehicles.

We decided to use the modular ambulance vehicle, as they are the best available today. They have large tanks for wall oxygen and wall suction, full telemetry capabilities, can comfortably transport three paramedics and up to three patients, and are easily repaired. We continue to use Life-Pak IV defibrillators. The Life-Pak V is a newer model and is a great deal lighter, but this feature has made it an easy item to steal. Systems that have used the Life-Pak V have encountered a much higher theft rate than we have. When setting up a mobile intensive care system, choose equipment which suits your particular needs. Of importance is the ease with which repairs can be performed when equipment malfunctions.

We stock a normal complement of emergency parenteral drugs on the ambulances. This was agreed to by a city-wide group of experts in Emergency Medicine. Our philosophy has been that the paramedic is an extension of the physician, rather than a mobile physician in his own right. The latter represents a definite philosophical trend in some areas of the country. The paramedics can only give drugs in our system when ordered by a physician or, in the event of a communications breakdown, when a patient's life is acutely threatened. Even under these circumstances, the utilization of parental drugs is dictated by written standard operating procedures.

We designed the system in this city to be totally advanced life support. There were to be no municipal ambulances run as basic life support vehicles. The only basic life support vehicles would be operated by the private ambulance sector. We also chose to have the advanced life support vehicles transport *all* patients to whom they were called, regardless of whether a true emergency requiring advanced life support was involved or not. This has been written into city ordinance.

There are some problems with such a system. The first is that wear and tear on the vehicles is considerable. Over a year's period of time, we needed to replace a vehicle every 2 weeks on the average. Each vehicle costs approximately $22,000, so this is quite expensive. Other major expenses are insurance, fuel, and repair expenses. One possible solution to vehicle wear is to have advanced life support vehicles respond to all calls, but to have them transport only those cases which definitely require their skills and to call on private ambulances to transport the other cases. Software is another major area of oeprational expense. Software includes the personnel operating the system, which includes paramedics and EMT-As. The total cost including men, equipment, maintenance, insurance, and fuel is between $200,000 and $250,000 per year.

The medical director should be a physician who practices emergency medicine full time, has been trained in the field and has a good working knowledge of it. He should preferably not be a medical specialist who has not worked in the field of Emergency Medicine. A physician whose expertise is limited to treating cardiac emergencies is not an appropriate choice for medical director of a comprehensive mobile intensive care system. This is because the mobile intensive care case load consists of only one-third cardiac-related runs.

Another important person is the Mobile Intensive Care Coordinator. This individual is responsible for paramedic evaluation and continuing paramedic education. This is an important job, and our region could probably use two such individuals, rather than just one. Operation of a paramedic training program boosts costs even further. It costs us about $900 to train a paramedic student using the 390 hour national curriculum.

Another aspect of software is a set of standing orders to serve as part of paramedic operating procedures. This will allow paramedics to perform critical treatment in the event of their inability to contact a physician due to radio malfunction or other problems. Our standing orders include those for treating suspected myocardial infarction, various arrhythmias, acute pulmonary edema, upper airway obstruction, shock and other emergencies. Our standing orders are always under revision and we are developing new ones currently in the areas of drug overdose and anaphylaxis.

We have found that roughly 20% of the time that a paramedic is called upon to use the skills in which he has been specifically trained, the patient has had successful results. There is another 15-20% of cases in which paramedic intervention *probably* made a difference in the outcome of a case. I am aware of only one gross error resulting from paramedic treatment in our experience, and this did not involve any mortality. Since patients already have intravenous lines started upon arrival in the emergency department, they are in better clinical condition, and work load on the physician and nursing staff is decreased. We are quite happy with the results, and believe that they more than justify the existence and expense of the system.

CHAPTER 4

PARAMEDIC TRAINING

Carole Swinehart, B.S.N.

There are two types of ambulance personnel in the field currently. The first is the emergency medical technician (ambulance) — the EMT-I. The EMT-I has completed an 81 hour training hourse in basic rescue and emergency care procedures developed by the American College of Surgeons. The EMT-I can splint, bandage, and perform basic cardiopulmonary resuscitation, but does not start intravenous lines, defibrillate, or perform more sophisticated procedures. These functions are carried out by the EMT-II. Before discussing the training of these more advanced Emergency Medical Technicians, it is useful to detail the history of their development in this country.

In 1966 the first mobile paramedic program was developed by Dr. Pantridge in Northern Ireland. The aim of this program was to provide paramedic services by people trained in coronary care. Dr. Grace instituted this system at St. Vincent's Hospital in New York several years later. The New York program was not staffed with paramedics, but it provided a model for other institutions to emulate. Programs began to develop in Seattle, Los Angeles, and Long Island,

New York. Each of these were locally planned. Each area developed its own paramedic training program to meet its own needs.

Illinois was the first state to coordinate local programs into a state-wide system. In 1973, the state established minimum standards for the training of paramedics. The recommended core curriculum consisted of 120 hours of instruction. Of these, 72 hours were to be didactic instruction and 48 were clinical training. Eighteen local programs developed in this state. The programs followed state guidelines but were different from each other, just as the programs in various areas of the country were.

Three years ago the National Academy of Sciences and the National Research Council formed a task force to develop national standards for paramedic performance and recommend a standardized curriculum. At the time, curricula for paramedic training varied from those requiring 80 hours of instruction of those requiring 1200 hours. In 1975 a grant was given to the University of Pittsburgh to develop a national curriculum for paramedic training. This has been done and the curriculum has been endorsed by the Federal Departments of Transportation; Health, Education and Welfare; and Labor. The assessment standards of the National Association of Science/National Research Council task force have been incorporated into the curriculum.

The national curriculum is a 390 hour program and is modular in design. It consists of 270 hours of didactic instruction and 120 hours of clinical training. The curriculum provides some educational guidelines to programs developing an M.I.C.U. system. According to these guidelines, the sponsoring institution should be an accredited post-secondary education institution which is affiliated with an accredited hospital or medical center. The primary supervisory person, the medical director, should have extensive critical care or emergency care experience. He should have ambulance experience and be capable of performing all of the skills the paramedics will be expected to do. In addition, he should have experience in instructing students at the educational level of paramedics, who have rather different backgrounds from medical students. Although most have a high school education, their orientation and comprehension is different from that of many other medical personnel.

The course coordinator — this is my role at the University of Chicago — oversees the operation of the program, acts as liaison between the students and the medical director, and is generally responsible for keeping the program functioning well. She is responsible for selecting the students, instructors, and training aids as well as scheduling the classes. She should also assist in the instruction.

Students must be at least 18 years old. They must have a high school education or its equivalent. They must have one years' active experience as an EMT-I, and be affiliated with a unit that is or will soon be functioning in advanced life support. The instructors may be doctors, nurses, or personnel engaged in allied health professions. The recommended class size is 20 students. The first class I taught contained 23 students, following attrition. This is a very manageable size for such a group. It is recommended that there not be more than two students in the clinical area at one time.

The course of instruction is divided into 15 modules, each of which is a complete and self-contained package. The modular design permits individual programs to develop in accord with local needs and resources. For example, there may be areas which do not have a cardiac patient load, but rather deal with a great deal of trauma. Such areas would not need to devote as much time to teaching of cardiac arrythmias, but could utilize the time training their students in the recognition of shock, management of fluid balance, and other subjects which are more trauma-related. Certain modules can, therefore, be deleted in accordance with a program's needs.

The program is based on the premise that the paramedic is physician-directed. The paramedic must, nevertheless, be a highly capable practitioner. He needs to possess certain skills in order to practice efficiently and also enough content knowledge to understand and the pathophysiology of disease. He must develop good judgement and be able to function safely in the event of his inability to contact a physician.

Each module contains both skill and knowledge objectives, which are to be applied in the clinical training areas. Clinical areas that are defined by the curriculum are the emergency department, intensive care and coronary care units, operating and recovery rooms, labor and delivery suites, and the psychiatric ward. Work on a pediatric unit and with an intravenous team is also recommended.

Going through each individual module, I perceive areas which may present potential problems to some programs. There are were some difficulties which we encountered which I am sure were not unique to our program. Module I concerns the Emergency Medical Technician, his role, responsibilities and training. This is an introductory module. It identifies the responsibilities the paramedic has in the field. The concept of medical ethics is introduced. The laws which permit paramedic functioning are detailed. In our state, the law provides the paramedic immunity from medical-legal action unless there has been willful neglect on his part. The module also recommends discussion of death, dying and methods of coping with the emotions which these subjects raise. I believe this point in the training period to be too early for discussion of the subject of death. I recommend reviewing it at a later time in the course.

Module II concerns human systems and patient assessment. It includes an overview of the anatomy and physiology of each system of the body. There is also an introduction to medical terminology, including the teaching of roots, suffixes and prefixes. The module contains the procedure for patient assessment. This includes the history, physical examination, and the transfer of information to the supervising physician. It takes longer than one module to learn these concepts and there is need to re-emphasize patient assessment continuously over the duration of the training program. The history includes the chief complaint, the history pertinent to this, and the past medical history. The physical examination consists of the "abc's" of resuscitation — airway, breathing and circulation — as well as control of hemorrhage. The secondary survey includes inspection and palpation of the head and neck, inspection of the chest, auscultation of the lungs and heart sounds, inspection, auscultation, and palpation of the abdomen, and inspection and palpation of the extremeties. The student must also evaluate the neurological status and the neuromuscular function of the patient as part of the secondary survey.

For purposes of instruction of patient assessment, we divided patients into four groups: 1) The conscious patient; 2) The comatose patient; 3) The trauma patient; and 4) Those whose medical problem is not trauma-related. We developed a form for the students to use to record findings during their examinations of patients in the Emergency Department. This forced them to record the vital signs as well as any findings they thought to be present. Forcing the students to record their findings made them examine the patient more closely. We deleted the teaching of a number of physical findings that we thought not to be applicable to the students' work experience. We taught the first and second heart sounds only, not gallop rhythms, since third and fourth heart sounds will generally not be audible during ambulance transport. For the same reason we did not stress the different types of bowel sounds. We did not ask the students to be able to differentiate between rales and rhonchi, as we thought this to constitute too fine a distinction. We should rather have students be able to detect the presence or absence of breath sounds, wheezing and stridor. We also chose to limit the teaching of percussion, as this technique is difficult to master and has limited use in the field. Patient assessment is a difficult concept to learn. It must be stressed adequately if the paramedic is to master it.

Module III deals with shock and fluid therapy. The concepts of electrolytes, tonicity and electrolyte solutions are introduced. Emphasis is placed on the manifestations of fluid and electrolyte imbalance. The causes, symptoms, and signs of shock as well as intravenous technique are taught in this module. This is what the students consider a "heavy module". The concepts contained are difficult for them to understand. It has taken them rather a long time to begin to apply the basic concepts of fluid, electrolyte and acid-base disturbances.

The students learn to start intravenous lines. They first learn to do this by starting them on each other. This constitutes a lesson to them of what patients go through when venipuncture is performed and gives them an appreciation of proper technique.

The student is more likely to try hard to start the intravenous line correctly on the first attempt when the person he is practicing on is someone he knows and also someone who is going to repeat the process in kind. The students improve their IV technique greatly during the course, although proper technique needs to be continually emphasized. The recognition of subcutaneous infiltration was something which they were particularly slow to learn.

The curriculum offers subclavian, internal and external jugular venipuncture as optional skills to be taught. Some systems have taught such skills to their paramedics, but we chose not to. We have elected to confine the students' intravenous sites to the arms.

Module IV concerns general pharmacology. In this module the students are taught the classification of drugs as well as their effects, indications, contraindications, dosages, and side effects. They are also taught the calculation of dosages and the metric system. They learn to administer medication intravenously, intramuscularly, and subcutaneously. The autonomic nervous system is taught, along with its component sympathetic and parasympathetic systems. This was another difficult module for many of the students. Some of the students' mathematics background was not sufficient for some of the calculations required. They therefore needed a great deal of coaching to accomplish the work required.

This module also involved a great deal of homework for the students. They had to prepare index cards listing the uses, therapeutic effects, indications, contraindications, and side effects of the major drugs they would use. They were also required to familiarize themselves with common oral medications, ones that ambulance patients might be taking. These include oral hypoglycemic agents, minor tranquilizers, antihypertensive agents, diuretics, and others.

Module V deals with the respiratory system. It begins with a discussion of anatomy and physiology of the respiratory system and the assessment of the patient with respiratory disease. Pathophysiology was presented for upper airway obstruction, chronic pulmonary disease, respiratory arrest, pulmonary edema, toxic inhalation, near-drowning, thoracic trauma, the hyperventilation syndrome, and pulmonary embolus. The techniques of treatment taught were oxygen therapy, use of a bag and mask, suctioning and endotrachial intubation. Intubation is a mandatory skill as described by this module, but I personally have some reservations about requiring each paramedic to be able to perform the procedure.

Module VI deals with the cardiovascular system. Anatomy and physiology of the system is taught. Some of this was rather difficult for the students to grasp — the action potential of cardiac cells, for example. The pathophysiology of coronary artery disease, acute myocardial infarction, cardiogenic shock, syncope, and cardiac trauma were presented. Pulmonary edema as a disease entity was repeated at this time. The interpretation and treatment of arrythmias was taught. The students were taught the performance of basic cardiopulmonary resuscitation, defibrillation, and application of cardiac monitors. In addition to these skills, the students were also taught performance of carotid sinus massage, but were given strict guidelines for its use. They were also taught the technique of intracardiac injection. We knew that there would be times when they would have no other means of administering medications to arrested patients, so this technique is a valuable one for them. This module is quite important and its content takes a long time to teach. Certain areas might even need to be repeated at a later time. The diagnosis of heart block, for example, is one with which many of the students had difficulty. Passage of a trasvenous pacemaker and phlebotomy were listed as optional skills in this module but we elected not to teach them.

The central nervous system is the subject of Module VII. Similar to the other modules, it begins with instruction of the anatomy and physiology of the central nervous system. Traumatic disorders, seizures, and strokes were among the entities discussed. The management of the unconscious patient is taught as part of this module. Techniques of spinal immobilization were introduced. We stressed the value of repeated evaluation of comatose or neurologically impaired patients.

Module VIII deals with soft tissue injury and begins with the anatomy and physiology of the integument. The injuries taught included abrasions, amputations, burns, and patients with impaled objects. The students were taught the types and degrees of burns as well as eye, face, neck, and abdominal injury. The skills that were taught as part of this module were those of hemorrhage control, splinting, and bandaging. Nasal packing is an optional skill suggested by the curriculum for this module, but we thought that this was not too practical and did not teach it. We were also not too specific in differentiating various types of eye injuries that we thought the paramedics could not specifically treat, such as recognition of retinal detachment.

Module IX discusses the musculoskeletal system, specifically the assessment and management of sprains, strains, dislocations and fractures. The use of traction splints, air splints, board and aluminum splints was taught. This module was readily learned by most of the students.

Module X deals with medical emergencies and it

includes identification and management of diabetic
ketoacidoses, hypoglycemia, anaphylaxis, heat
stroke, hypothermia, frostbite, poisoning, alcoholism,
overdose, the acute abdomen, and management of the
geriatric patient. A good many subjects were lumped
into this module because they did not fit neatly into
any other. With regard to the acute abdomen, we
stressed the treatment of shock when this is present
and did not dwell on the individual entities which
cause abdominal catastrophies. After all, we did not
want the paramedics to think that a detailed examination of the abdomen on their part superceded the
necessity to treat shock.

Module XI covers obstetric and gynecologic emergencies. Subjects ranging from rape and pelvic inflam-
matory disease to placenta previa, abruptio placen-
tae, prolapsed umbilical cord, and postpartum hemorrhage were taught. The students spent time in the
delivery suits under the direction of the obstetric
residents. This was a most meaningful rotation for the
students. Many of them said that their only prior
training in obstetrical delivery had been through
rather dated training films. All the students felt much
more secure in managing a delivery in the field. Even
students who had performed several deliveries found
the experience useful.

Module XII is entitled "pediatrics and neonatal
transport". In this section the paramedics were
taught problems unique to children. We discussed the
pathophysiology of pediatric emergencies, such as
asthma, bronchiolitis, croup, epiglottitis, the sudden
infant death syndrome, seizures, and child abuse. The
differential diagnosis among some of these entities
will be quite difficult for the paramedic to make. I am
not sure how valid it was to have taught them these
entities to the depth the curriculum suggested. The

students were taught how to perform cardiopulmonary resuscitation on infants and they were given some instructions on starting pediatric intravenous lines.

Module XIII covers the management of emotionally disturbed and psychiatric patients. The students were taught procedures in handling suicidal, violent, paranoid, depressed, and hysterical patients. Few of the psychiatric patients that our emergency department comes into contact with come to the hospital via the ambulance system. Therefore, the exposure our paramedics have to psychotic patients is limited, and we did not go into great depth in training them to identify specific psychotic problems.

The last two modules deal with telemetry communications and rescue technique. Extrication and rescue techniques were taught with the use of wrecked cars. We spent some time discussing electrical and toxic gas hazards. The rescue techniques which individual communities will require its paramedics to know will depend upon the geographical conditions within those areas. The modules should be modified to suit these conditions. The teaching of triage was not contained in any of the modules, so we taught it at this point. We wanted the students to be able to identify emergency conditions and establish priorities of treatment since this is a role for the paramedic at the site of a disaster.

After observing the students in the clinical area for six months, I find that they are very skill-oriented. They worry more about their inability to start intravenous lines than concerning themselves with total patient orientation. They are often so obsessed with skills that they fail to communicate well with patients. The curriculum advises that the students are to be kept in the clinical areas until they have experience with all the entities taught to them in class. Obviously this is an unrealistic goal.

The modular-design curriculum seems very reasonable on the surface. However, it is based on the assumption that all those who enter the program can complete the course if given enough time. This is also unrealistic. No matter how hard you work with some students they will not be able to grasp the content of some of the more difficult modules. We have set an arbitrary deadline of six months for completion of the course. I would personally make the course longer, but the line must be drawn somewhere. A certain tedium develops when the students see and listen to you three nights a week for months on end. Their basic time requirement is 20 to 25 hours per week. This includes 9 lecture hours, 8 clinical training hours and time devoted to homework.

One problem we encountered was the lack of a good basic textbook. As a result, a number of readings had to be assigned from other sources. Hopefully, a textbook will be developed soon to rectify this lack.

Our philosophy holds the paramedic should function primarily in stabilization and transport of patients. As such, the amount of theoretical knowledge required in some areas of the curriculum is excessive. It is inappropriate to overtrain paramedics and expect them to perform diagnostic functions rather than stabilization and transport. It is also unreasonable to expect most paramedics to master the content of the curriculum as well as the skills required. Therefore, it is important to modify the curriculum to conform it to your idea of what a paramedic is and what you expect him to do.

CHAPTER 5

MEDICAL-LEGAL POLICIES, PROCEDURES, LIABILITY, & CONSENT

Marshall B. Segal, M.D., J.D.

A legal maxim to be followed in Emergency Medicine is: "When in doubt, treat". Another way of stating this is: "If you are going to make a mistake, err on the side of life, err on the side of treatment." Try to be a dedicated doctor or nurse at all times. That is a very defensible position. Do not try to think like an attorney. Think like a physician who is concerned about his patient. You will have the sympathy and understanding of the jury only if you have appeared to act in the best interests of your patient. If you appear to be acting to cover medical-legal liabilities you will not appear very defensible to a court of law no matter what the law involved. Furthermore, never act against your own medical judgement. This is like betting against yourself, and puts you in a "no win" situation. If something goes wrong after you have abandoned the course dictated by your medical judgement, your defense will be difficult.

There are several circumstances under which one may be sued. These are contractual liability, intententional torts, negligence (which is an unintentional tort), and statute. Contractual liability is rarely an Emergency Department problem. A contractual agreement in medicine might exist between an obstetrician and a pregnant patient. For example: A doctor agrees to care for a patient and deliver her child when the time arrives. He can reasonably expect her to deliver 240 to 300 days following conception. Suppose then that when she goes into labor her doctor is unavailable to perform the delivery and has made no arrangement in his absence for coverage of his practice. The woman's baby suffers morbidity as the result of this absence. She may sue on the grounds that the doctor abandoned her in violation of his contractual obligation.

Contracts are seldom in force in the Emergency Department. This is because in emergency situations doctors and patients do not sit down at a table and negotiate terms for treatment. The fact that immediate treatment is required makes this unreasonable. The small print on the back of the Emergency Department treatment form signed by the patient is merely an adhesion contract. The courts will not give it much weight as a negotiated contract. A patient who believes himself to be in desperate need of treatment will sign anything put in front of him.

There are, however, a few contractual situations that do apply to Emergency Medicine. If you are involved in health maintenance organizations or prepaid medical programs such as Group Health in Seattle or Kaiser Permanente in California, provision of emergency medical services is an integral contractual part of such plans. Another example involves hospitals which have agreements with the government to take care of Jones Act Workers. These are maritime workers on American vessels on the Mississippi River and the Great Lakes. If your hospital has an agreement with the government to care for such workers, the Emergency Department is bound to offer services for such workers *on a contractual basis.* Agreements of this sort may become more common in future years as American medicine becomes more organized along such lines, but contractual abandonment is not a large part of Emergency Department liability at the present.

A major intentional tort is assault and battery. The concept of patient consent revolves around this tort. I am referring to civil assault and battery, rather than the criminal action. The terms "assault" and "battery" are often misunderstood. Assault is an attempt to touch another person, battery is the actual touching. Should I swing at you and miss, that is assault; if I actually hit you, that is battery. There are conditions, however. Touching without consent does not constitute battery when it is socially acceptable or when consent has been attained. Obtaining a patient's consent for treatment is done to ensure that what follows is not an unconsented touching of the patient by the physician. Bodily integrity is very important in this culture. It has political, social, sexual, and emotional connotations. We are a much more "non-touch" culture than are many European cultures.

In order for assault and battery to occur, there must be an intent state of mind. Suppose, for example, that I am an epileptic having a seizure, and you are attempting to treat me. My arm flies out, breaking your jaw. I am not guilty of assault and battery because I am not conscious.[1]

Another intentional tort is false imprisonment. The criminal equivalent is undue restraint. No physician or nurse has ever had to pay a dollar for using therapeutic restraint rationally. You must, of course, use nondeadly force to restrain patients. You cannot, for example, shoot a patient who is trying to leave your department in order to get him to stay for treatment. On the other hand, suppose you are forced to restrain a psychotic patient who is threatening to climb to the roof of your hospital. You are restraining him because you are afraid he will throw himself off the roof. If in the course of restraining him you accidentally break his arm, you are not liable. Weighing that injury against the possible harm which could have come to him had he jumped, you and the patient come out ahead. You were in effect trying to safe the patient's life. Whenever you are forced to restrain a patient it is reasonable to document the reasons and the fact that this is being done on medical grounds.

Bodily restraint may be more subtle than this. For example, if a patient threatens to leave your department without paying his bill, you cannot threaten to keep his clothes or false teeth locked up until he agrees to do so. He may claim that he does not want to be seen in the street without his false teeth and that your refusal to return them constitutes false imprisonment. This is not to say that you cannot go through due process to obtain payment, but you may not employ bodily restraint.

Libel and slander are the intentional torts of defamation. Slander is speaking badly about someone such that it hurts him, especially professionally. Libel is the written equivalent. As long as your communications remain within a confidential privileged relationship, libel and slander cannot justifiably be brought against you. Let's suppose my nurse says to me that a particular doctor in my Emergency Department is incompetent and should not be entrusted with the care of an Emergency Department patient. Even if this doctor had a tape recording of this conversation, he would not have grounds to sue that nurse for slander. He would lose a suit against the nurse because the statements were made while she was engaged in a privileged conversation. You are entitled to speak to your administrator, colleagues, and associates within the framework of organized patient care and not be held for slander. The courts realize that if they allowed a suit of this sort, communication to improve patient care would be impeded. However, if you go to a cocktail party and repeat the same conversation or voice similar opinions, you may be liable. You must keep communications within professional channels in order for them to have immunity from action for libel or slander. In addition, in the American system, truth is an absolute defense against slander. This is not so under the British system, where even a truthful statement — if it is damaging — could possibly be actionable.

Confidentiality is a subject related to privileged communication. It is the physician's ethical responsibility to remain silent about his patients. Breach of such responsibility is quite actionable. *Privilege* is the power to resist being forced to speak about a patient in a court or court-like proceeding. Doctor-patient privilege is not contained in the United States Constitution, but rather covered by state statute. As such it is quite variable from state to state. For example, in Florida, there is a psychiatrist-patient privilege, but not a physician-patient privilege. Remember that it is the patient, not the physician, who possesses the privilege. From an ethical standpoint however, the physician should exercise the privilege until the patient allows the physician to testify. In the state of Illinois there are some exceptions to the privilege law. These include treason, cases involving an attack on the President of the United States, and cases of homicide. Obviously, these cases are of an unusual sort.

The tort of outrage is not frequently invoked. It provides that even if damages cannot be proved against a physician, he may be found liable if he has been outrageous in his conduct. The well-known suits against the psychiatrists who had sexual relationships with their patients are cases in point. It is difficult for a patient to set specific damages resulting from sexual intercourse. The patient's attorneys, therefore, may claim that this is a tort of outrage — a violation of the responsibilities of the office of the physician. Punitive damages may be asked on the basis of this violation of trust. The nurse who persuades the old man under her care to sign her into his will will not inherit the estate. In addition, the family may sue her for punitive damages.

Negligence is an unintentional tort. The intent to be clumsy is not necessary for negligence to occur. What are the elements of negligence? First of all, there must be a *duty* to the patient. If you encounter an injured patient on the street you have no legal duty to assist him. That is, unless you are in the state of Vermont, where an affirmative duty is created by specific law. However, when a patient crosses the physical and metaphysical boundary between the street and the Emergency Department, the duty attaches. If a drunk staggers into your Emergency Department you cannot allow him to stagger back out, no matter how tempting this might seem. He staggers out with your duty towards him attached. This is not the case in a private physician's office. The private physician may refuse to allow a patient into his practice. An Emergency Department, however, may not turn a patient away without at least a reasonable medical evaluation.

The next element of negligence is a deviation from his duty. The deviation from duty is determined by standards. In addition to deviation from standards, there must be harm or *damages.* Bridging these two, there must be a *proximate cause,* some sort of relationship between the deviation and the harm. Even when there is deviation from well-accepted medical standards, the physician is not liable if these do not constitute the reason for damages in a particular case.

How are standards established in court? The simplest way is through expert testimony. Let us assume that I observed an automobile accident involving two cars. I can be called to court to testify in this accident case. I can be asked questions concerning the details of the accident, the weather, visibility, and the road conditions. I cannot properly be asked whether in my opinion either of the cars was going too fast for existing conditions unless a foundation has been laid to the effect that I am an expert professional driver. Not until such a foundation has been laid may I answer an opinion question of this sort. These principles apply to medical cases as well. If a physician is being questioned as a witness in a medical litigation, he cannot be asked for his opinion unless a foundation has been laid that he is an expert witness. Incidentally, the physician is entitled to an expert witness fee for such expert testimony.

Alternatively, text books may be used to establish professional standards in many states. Sometimes legal statutes establish standards. For example, there is a law in Illinois which provides that all hospitalized women must have a Pap smear performed. If this is not done and a suit is brought up as a result, the patient's attorney does not need to establish standards on the basis of custom. The standards are set down by the statute.

A hospital or department's own policy and procedure protocols may also be used to determine standards. It is therefore necessary to establish reasonable policies and assure that protocols are worded in a reasonable manner. For example, details which are likely to change frequently — such as drug dosages — should not be written into standing protocols.

There isn't very much case law in Emergency Medicine. A case which results in a human tragedy tonight will not come to the appellate court level for six years. In 1983 the court may judge a physician's actions in the 1983 standards, despite the fact that the case occurred in 1977. In addition, the "community standards" defense is not a very strong one. If a department does not have a defibrillator, it doesn't matter that the other hospital emergency rooms in an area do not, either. There are many community hospitals who do have defibrillators in their Emergency Departments. The department in question will be held to a national standard of good patient care.

The emergency treatment of minors involves certain legal aspects of consent. The Illinois state law, 91 § 18.3 has this to say about minors:

> Situations where consent need not be obtained: Where a hospital or a physician, licensed to practice medicine or surgery, renders emergency treatment or first aid or a licensed dentist renders emergency dental treatment to a minor, consent of the minor's parent or legal guardian need not be obtained if, in the sole opinion of the physician, dentist or hospital, the obtaining of consent is not reasonably feasible under the circumstances without adversely affecting the condition of such minor's health.

The test is "adversely affect" not "life or limb". This implies that every minor who presents as a patient in the emergency setting is entitled to a history and physical examination whether the parents are present or not. The best way to be sued is to fail to do this. There is no other way to tell whether a minor patient's health will be adversely affected by withholding treatment.

What if, after examining a minor, you discover he might have a ruptured spleen? Can you do a peritoneal lavage without the parent's consent? Certainly. You can perform invasive procedures up to and including an operation in the absence of the parents. To house a minor is to take on the responsibility to care for that minor. To do otherwise is to violate the "dedicated doctor" rule.

The parents are also required to pay for such care whether they consented to it or not. The child and his parents are liable for any contract the minor makes for the necessities of life. This is basic law. What if the child refuses treatment? I would listen to what he has to say. He might, for example, be saying, "I am a hemophiliac and can't go into surgery." That is, he may be offering significant medical information. He may not, however, refuse treatment. He is chronologically incompetent; that's what being a minor means. Since he has no legal competence to agree to treatment, he lacks the legal competence to refuse as well.

What if the minor is unconscious? Since he was incompetent when he was conscious, he is no more competent when unconscious. You may procede with treatment. What if both parents are present and one agrees to treatment and the other does not? The law opts for treatment in most cases. Most state statutes say, "Any parent can consent to treatment." Notice that the singular form is used in order to deal with just such an eventuality. Whenever a physician opts for treatment he is a reasonable person trying to maintain a child's health. If he favors the negative reply, and the individual who refused permission to treat the minor turns out not to even be the child's parent, the physician is in real trouble.

What if the child is brought in for treatment by a babysitter? Go ahead and treat the child, but get the babysitter's signature on the chart. This person has been entrusted with the care of the child and this includes bringing the child to the hospital should this be necessary. The babysitter has the implied authority of the parents to consent to treatment.

What about both parents being present and *both* refusing treatment for the child? In this case the test of whether to treat changes. It is no longer, "adverse affect" but rather "substantial injury", or "life or limb". The child abuse laws come into play here as well. These state that a physician can admit and house a minor in a hospital. The implication is that a minor can be treated under these laws, since the legislative intent was obviously to protect the minor.

Consent is somewhat different when adult patients are involved. There are three types of consent: expressed, implied, and constructive. Expressed consent is consent directly articulated by the patient. Implied consent is that consent suggested by a situation: A man staggers into the Emergency Department complaining of crushing substernal chest pain. No further formality need be gone through to obtain expressed consent for treatment of this patient. Consent has been implied by the presentation to the hospital. Signatures on a consent form are nice for documentation. They are very important for third party benefit payments. Actual consent in cases such as this one is, however, implied, and a signature is not essential when inconvenient. Constructive consent is sometimes mislabeled "implied". Constructive consent is the child of necessity imposed by society in cases where it is not impossible to obtain expressed consent. The unconscious victim of an automobile accident does not have to wake up and say, "Yes, you can do a craniotomy to save my life," before this can be done. The situation is very much like that encountered with a minor. The housing of an unconscious patient implies the care of such a patient. Neurosurgeons have asked me what to do under similar circumstances when they have been able to contact the spouse of an unconscious patient. I have uniformly advised them to operate. What if for some reason the spouse had refused to consent to treatment? Should treatment be withheld? Certainly not. A "significant-other" cannot commit a spouse to death. Always act to preserve life. Consider that you have the legal responsibility to maintain the *status quo*. Keep the patient viable until the judge can decide the issue. A dead patient is not a viable legal issue.

Let us take the instance of the patient in the Emergency Department who needs blood, but says, "Don't give me blood, I am a Jehovah's witness." In my opinion no one in my Emergency Department who requires blood is competent to make the decision of whether or not it should be used. He is either in shock, anemic, hypoxemic, or a combination of these., What if the patient is really an Orthodox Jew who earlier that evening had watched a television program about Jehovah's Witnesses? The shock state induced by his bleeding duodenal ulcer may have caused him to be confused. Don't believe that this cannot happen. Certainly, a suit can be brought against a physician for wrongly administrating blood to a competent, conscious, alert Jehovah's Witness, but it will not be of the same seriousness as that suit brought by the family of the patient from whom life-saving treatment was mistakenly withheld.

There is the case of the unconscious patient who requires transfusion whose spouse tells you that he is a Jehovah's Witness. If he were awake, this man might tell you "You know, I'm religious, Doctor, but I am not *that* religious. Go ahead and give me the blood." On the other hand, perhaps the patient is not a Jehovah's Witness. but the wife in all good faith has always wanted him to be, for the sake of his eternal salvation. Critically needed therapy might be withheld from this man on the basis of a false assumption. I don't take religious principles lightly but only use these cases to illustrate that the first mistake you can make is to let a patient die when it was within your power to prevent his death.

Diminished capacity cases are the most difficult. Culturally, a blood sugar of 600 is easy to deal with, but a blood alcohol level of the same magnitude is not. Certainly the doctor knows that he must retain the diabetic in ketoacidosis but must he also retain and treat the alcoholic who wants to elope. The Alcoholic Treatment Acts tell us that we must, but they do not say for how long. The drunk driving statutes in Wisconsin and Illinois define drunkeness above blood alcohol levels of 150 and 100 mg%, respectively. However, chronic alcoholics develop a tremendous tolerance and levels of this magnitude would not render them intoxicated. When treating an alcoholic, you should have at least one blood alcohol level drawn so that you can be sure you are dealing with a case of alcohol intoxication and not diminished capacity on some other basis, for instance a subdural hematoma or meningitis. You should discharge the patient from your care when you deem it medically appropriate, and you should document this fact on the patient's medical record.

The problems of diminished capacity are magnified in the case of narcotic abusers. We have an excellent drug in naloxone to reverse the effects of narcotics. Unfortunately, it is a short-acting drug. Many an addict has been awakened from coma by the injection of naloxone, who then wants to go downtown to transact some business. How do you explain to him that he might lapse back into unconsciousness and even respiratory arrest once he leaves your care? He might not believe that he was unconscious at all! In this case you must obviously retain a patient for therapeutic purposes.

If your hospital has received funds under the Hill-Burton Act, one of the federal code of regulations states that you may not refuse treatment to an alcoholic or drug addict merely on the grounds that he is one of these. You may not say for example, "We will not do your appendectomy because you are an alcoholic. Having alcoholics around here ruins our environment". This does not mean that you must maintain alcoholic treatment centers, nor do you need to perform the function of such a center in your hospital. You cannot, however, refuse medical treatment to alcoholics or addicts merely on the basis of their substance abuse.

What about the critically ill patient who refuses treatment? My real question is "How many critically ill patients are really competent?" If you believe a patient to be incompetent on the basis of his illness you should document this fact on the medical record and procede with treatment. In the Emergency Department you can go as far as you need to as long as you document your professional opinion based on observed medical signs and symptoms that the patient is incompetent.

If a patient wants to leave your care and you believe that he is competent, have him sign the against medical advice (AMA) form and let him go. No one *has* to sign an AMA form before being permitted to leave. Restraining someone until he has signed such form constitutes false imprisonment. If the patient refuses to sign, merely inform him and state in the record that he has been informed that he may become very ill or die if he does not agree to treatment. This covers most of the adverse possibilities. I prefer that the nurse on the case also place this into her nurses' notes. Are three signatures required on an against medical advice form? This may be your hospital policy and the signatures may be useful to have in case of eventual litigation, but it is not a legal necessity.

Should you include a patient's expletives in a medical record? I do when I deem it necessary and I have educated my medical records department to accept such documentation. The nasty person who would use such language is the same sort who will sue you later. It is best to document the full character of his behavior.

What about informed consent for major procedures? Major procedures in the Emergency Department are generally of two types: One is a sort that is done by non-emergency specialists and the other by the emergency physician. Obtaining consent in the former case is the responsibility of the specialist performing the procedure. In the latter case, the situation is usually very urgent and informed consent is not needed. Culdocentesis in the event of ectopic pregnancy or peritoneal lavage in the face of abdominal trauma are procedures done when time is of vital essence. The situations which call for the performance of these procedures often do not allow for the luxury of time for consent. It is appropriate to explain procedures to a patient if you have the time to do so. The more you explain, the more cooperative the patient is likely to be, but informed consent is not necessary when time spent obtaining it could be used to stabilize the patient's condition.

Can a blood alcohol level be drawn without a written consent form having been signed? Certainly. A blood alcohol level is just like an SGOT for medical purposes. Unless the patient specifically objects, regard it as part of your diagnostic regimen. You must not reveal the results of this test to anyone without proper authorization. A patient will not sue you for drawing a blood alcohol level, but he may sue you for revealing the results inappropriately.

Regarding procedures and policies, there are certain things which should be in all Emergency Department procedure books. This includes the text of certain laws. The Uniform Anatomical Gift Act is one which deals with obtaining donor organs for transplantation. Such a law has been passed in all fifty states. The abused and neglected children acts, emergency medical services act, and consent for treatment of venereal disease, prescription of contraceptive devices, and for treatment for abuse of controlled substances. The good samaritan laws for nurses and physicians should be present. Nursing practice acts should also be included. There has recently been legislation in some states regarding patient's rights to their records. It is useful to have copies of such statutes where they exist. Many states have laws describing alcoholic treatment guidelines or rape victim's treatment guidelines. These should also be included. Such laws and others applicable to emergency practice are useful to have at hand for purpose of reference.

FOOTNOTES

1. This conduct might, however, constitute negligence. For example, you may be able to prove that I am an irresponsible alcoholic and am prone to seizures if I drink too much. Therefore, by drinking excessively I am putting you in bodily jeopardy. You may therefore sue me for negligence, but not for assault and battery, because I did not have the intent to strike you.

2. *Stmt:* Illinois Statute 51 § 5.1 states: No physician or surgeon shall be permitted to disclose any information he may have acquired in attending any patient in a professional character, necessary to enable him professionally to serve such patient except only (1) in trials for homicide when the disclosure relates directly to the fact or immediate circumstances of the homicide, (2) in actions, civil or criminal, against the physician for malpractice, (3) with the expressed consent of the patient, or in case of his death or disability, of his personal representative or other person authorized to sue for personal injury or of the beneficiary of an insurance policy on his life, health, or physical condition, (4) in all civil suits brought by or against the patient, his personal representative, a beneficiary under a policy of insurance, or the executor or administrator of his estate wherein the patient's physical or mental condition is an issue, (5) upon an issue as to the criminal action where the charge is either murder by abortion, attempted abortion or abortion or (7) in actions, in civil or criminal, arising from the filing of a report in compliance with the "Abused and Neglected Child Reporting Act."

CHAPTER 6

MULTI-HOSPITAL MANAGEMENT

Thomas Cooper, M.D.

I am the medical director of an organization which contracts to provide Emergency Department coverage at hospitals throughout the United States. At the present time we provide coverage at 75 hospitals in 23 states. The hospitals vary in size from 30 to 750 beds and populations that these hospitals serve range from 3,000 to 150,000 people. Emergency Department patient volume at these hospitals also varies greatly. There are some hospitals which annually see 3,000 patients in their Emergency Departments, and some which see 40,000. We do not provide continuous 24-hour coverage at all these hospitals. At about a third, we provide weekend coverage.

The subject of multi-hospital management includes a number of concepts. These include the functions and services that a central management team provides to the physicians and hospitals it serves, continuing education for a diverse medical group, and financial management. Our organization has developed a system of audit criteria by disease categories which I will describe. A discussion of multi-hospital management also involves marketing; namely, how physicians and hospitals get in touch with us to provide and receive our services.

The first question that many physicians have about multi-hospital management is: What is the purpose of such an organization? There are a number of advantages to multi-hospital management. When a physician contracts with a hospital to provide emergency services, he becomes involved with certain individuals with specialized backgrounds. These include hospital administrators, many of whom now have masters degrees in the field. Such administrators have backgrounds in accounting, financial management, statistics, organizational behavior and administration, management of health care organizations and law. Most physicians have no training in these fields. What we have attempted to do in our organization is to have members on our staff who have the benefit of such backgrounds. Three of our staff have masters degrees in hospital administration as well as experience as hospital administrators. They maintain close relations with all our physicians and medical directors, aiding them in dealing with hospital administration. They also meet periodically with hospital management to evaluate the status of the contract, utilization of the Emergency Department, interaction with the remainder of the hospital, especially hospital areas which support the Emergency Department — such as the pharmacy, the laboratories, x-ray, and inhalation therapy. They review the statistical and financial breakdown of the Emergency Department and public relations programs as they relate to the Emergency Department. They also consult on facility design and equipment.

Our hospital administrators and medical directors also work with hospital administration in establishing staffing patterns for nurses and ancillary personnel working in the Emergency Department. They conduct and assist in the initiation and modification of written medical and administrative records. They perform an evaluation of the Emergency Department based on the current accreditation standards of the Joint Commission on Accreditation of Hospitals. They consult with all physicians in establishment and modification of group policies and practices.

One of their main functions is assisting our physicians in getting things done. The gap between doctors and administrators can be a philosophical one. Physicians are often end-oriented, while administrators are means-oriented, and the two groups sometimes have a difficult time finding a common ground. Our hospital administrators try to bridge this gap.

The hospital attorney is another individual with whom physicians must interface. Many hospitals now employ a full-time salaried in-house attorney. Since such an attorney works for the hospital exclusively, it is not difficult to see which side he will favor in any negotiations between hospital and physician. St. Louis University has established a combined degree program in hospital administration and law. The future will find hospital administrators becoming more sophisticated in legal matters. This will have to be true of physicians as well.

With this regard, our group has an attorney on its staff. He deals only in Emergency Department legal matters and is quite a crucial figure in our organization. He writes about 400 contracts per year, is involved in analyzing the malpractice insurance market, consults with hospital legal counsel in matters involving the Emergency Department, assists our physicians in personal legal matters and periodically reviews changes in the law which may affect our organization. He also is involved in negotiation procedures with hospitals.

The hospital comptroller is a difficult individual for many physicians to deal with because physicians have a difficult time relating to many comptrollers' apparent fixation on the bottom line. A hospital comptroller is charged with serving the hospital's interests rather than those of the physician. A member of our staff served as a hospital comptroller for a number of years prior to joining us. He is invaluable in any meetings we have with hospitals comptrollers. Specifically, he is involved in analyzing the past financial picture of an Emergency Department and projecting future financial indications. Such a financial picture includes analysis of patient volume, fee structure, billing and collection capabilities, third-party carriers and workman's compensation. He manages and coordinates our computerized physician payroll and works out the subleties of salaries, taxes, overage distribution and related details. He also assists in contract negotiations with hospitals. In instances in which we are involved where the hospital performs patient billing, he monitors the hospital billing and collection performance. He audits all statistical and financial information that the hospitals submit to us.

We have developed a billing system for our physicians. Our central computer terminal bills about thirty thousand patient accounts annually, and this is a service which we can offer to hospitals who are not interested in performing this function. Our comptroller establishes a relationship with third party carrier groups in each state in which we practice.

The administrator, attorney and comptroller are key individuals on our staff and represent important counterparts in the hospital administration. In addition, our office staff includes three physicians who do administrative work in addition to performing clinical duties. This arrangement assures that medical decisions are made by physicians rather than by non-medical personnel. These physicians become involved in evaluating any medical problems which may arise. They serve as liaison between our physicians and the medical staffs of hospitals where we have no medical director. These are the small hospitals where we provide the part-time coverage. One of these physicians typically works the first week of any new contract, in order to assure a smooth transition. All of them assist in our continuing medical education program.

We have an operations department which deals with scheduling, credentials, malpractice information and other related matters. It is a demanding job just to maintain up-to-date credentials information on 800 physicians. The part-time coverage hospitals provide particular problems, particularly those of last-minute changes in staffing. We maintain a 24-hour answering service with one administrative person and one physician on call to deal with any such problems which might arise.

By spreading the cost of these services over 75 hospitals, expenses can be minimized, but a quality of expertise maintained that befits a group working exclusively in Emergency Medicine. This is not to say that multi-hospital management is the only way to perform such services. It is one option though, and one which we have utilized very successfully. A doctor could certainly perform these tasks himself, if he had the inclination to do so. There are few physicians, however, who possess a sufficient grasp of all the above aspects of administration. In addition, very few physicians will be able to generate an income performing administrative work comparable to that which they could earn treating patients. The clinician must consider how much it costs him to devote the time needed to perform administrative tasks. If he enjoys negotiating contracts, recruiting physicians, and performing all the other tasks involved, he may be willing to sacrifice the financial gain he could have made working in the clinical area. Another consideration is whether a physician will be able to do the job as well as a trained professional. It is difficult for the physician to acquire the expertise required to carry out administrative functions as well as specialists in those areas.

Hiring such specialists on a part-time basis also poses problems. A major one is that part-time employees have work commitments elsewhere. Our attorney, comptroller, and hospital administrator are on our team. They are paid by our organization and have no outside employment. Their professional lives are dedicated to our organization. The multi-hospital administration approach takes advantage of the fact that there are many physicians who do not want to perform administrative work. They want to practice Emergency Medicine, but don't want to become involved in all the administrative aspects of patient care.

A major problem for both single and multi-hospital groups is that of recruiting qualified physicians. Our organization conducts recruiting programs to attract qualified Emergency Physicians on a national scale. About 3,000 physicians contact us each year. No full-time physician is taken on by our organization, however, without undergoing a personal interview, verification of his educational training, and checking of his references. We try to match the physician's professional needs to our specific employment opportunities. We continually communicate advancement opportunities within our organization to our physicians. Recruiting is an expensive procedure. It is also a time-consuming effort and the time involved in telephone conversation, interviewing physicians, traveling to meet physicians, verifying credentials and the other tasks involved in recruiting is considerable. Moreover, not all physicians have the expertise to adequately evaluate prospective physician associates. There are recruiting firms which will perform these functions, but these charge eight to ten thousand dollars per physician. Establishing a four-man group and recruiting all the physicians through a recruiting company might cost forty thousand dollars.

We are continually developing our continuing medical education department. We maintain libraries with current volumes pertaining to Emergency Medicine. We maintain audio educational materials for the benefit of physicians in every hospital with which we work. We issue a monthly publication of the newsletter type which is distributed to all our physicians, administrators, and other employees. It contains medical information as well as other news, such as feature information about our physicians and hospitals.

Communications is a key to an organization such as ours. When contracting with a particular hospital, one needs to know what is going on there, the staff and administration views on issues, and what problems there exist. We have established a formal communications vehicle. This is the medical director's report, which is filed every thirty days. The first item on such a report is education: How many hours of clinical instruction have been given to the nurses, the emergency medical technicians, the full-time emergency physicians? Instruction given to the remainder of the medical staff is also recorded. One of the reasons we collect this data is to indicate to a director the sort of activity he should be undertaking. Another section is devoted to public relations. This includes a listing of meetings with community groups, the purpose being to promote the Emergency Department. Such meetings introduce Emergency Medicine to the community, describe the problems we are involved in, our beliefs and goals. Another part of the report is concerned with medical staff obligations. Any existing problems with nursing and medical staffs are reported here.

Most of the time problems within a hospital can be satisfactorily handled on a local level. However, the above reports constitute documentation of problems that do exist. Such documentation is often demanded by hospital administrators who are called upon to deal with a particular problem — and rightly so.

If the hospital does our billing, there is crucial financial information we require in order to run our operation. This includes funds generated by the Emergency Department, the hospital's minimum guarantee, coverage over this minimum, the number of patients seen and the percentage of those patients seen by our physicians, etc. Each physician interviewing with us sees a statement prior to signing a contract. The information that these physicians receive includes much the same information: the patient fees, funds generated, malpractice insurance, our management fee, and other data.

One of our physicians is designated as the medical director at each hospital with which we have a contract. He receives extra compensation for his services. The way in which we manage our operation is not necessarily the only way to do this, but it is one way, and as I have indicated, we are quite successful at what we do. We have had to deal with complex staffing patterns and have been able to staff hospital emergency rooms successfully. For example, if a particular physician is unable — on short notice — to work his shift, we are able to provide backup service efficiently through the system described.

The Joint Commission on Accreditation of Hospitals has indicated that audit criteria by disease categories for emergency care must be established. We have expended some effort to do this, and have developed an audit for seven entities. These are sore throat, urinary tract infections, abdominal pain, chest pain, fractures, wounds, and lacerations. We feel that hospitals and Emergency Departments will be called upon to produce some criteria for an audit of medical care. It is, therefore, wiser for physicians to establish such criteria than to have them dictated by some governmental agency. We have itemized points of information that must appear on the medical record for each of the above entities to define the care delivered. These include the chief complaint, impression, and times the patient came in and was discharged. For the category of abdominal pain, it would also include the details of the abdominal, rectal, and pelvic exams, a description of the pain, and laboratory work. After establishing information criteria, we have the hospital medical records department perform a retrospective analysis of Emergency Departments charts of all patients who presented with abdominal pain, noting the above points. Patterns of care can then be observed, and if it is found that a physician in a particular area is delivering medical care on grounds greatly different than all others, the matter must be looked into. No one appreciates being scrutinized in this way, but, as I have indicated, I think that some form of audit is going to be imposed in the future. It is wise to establish a system for doing this before this occurs.

Of our 75 hospitals, only 6 had an established system of Emergency Department physician coverage prior to their agreement with us. Of the six, two had physician groups that were discontinuing service to those hospitals. We are very strong supporters of the ethic that if there is a pre-existing group providing Emergency Medical services at a hospital, that group should be made aware that we are negotiating with their hospital administration. Most administrations are agreeable to this policy, but a few are reluctant. If an administration is seeking alternate Emergency Department coverage, the Emergency physician currently performing that service has a right to know that this is occurring. This policy has gained us respect from administrations and has been appreciated by existing physician groups. We appreciate — though we do not always receive — the same courtesy in return. If we are interested in contracting with a hospital, but find that organized physician coverage already exists, we normally will not pursue our interest. The hospitals in which we seek to replace groups are generally those whose hospital administration has contacted us.

CHAPTER 7

THE EMERGENCY DEPARTMENT MEDICAL RECORD

Michael C. Tomlanovich, M.D.

Completion of medical records is a tedious but very necessary part of emergency medical practice. It is very time-consuming in most institutions and probably takes up one-fourth to one-third of a physician's actual patient care time. Filling out records is a task which many physicians perform less than optimally, and this can, in part, be related to the design of medical records. The components and requirements of medical record keeping are becoming more and more stringent. Many agencies are requiring better documentation. In fact, several states have passed laws in relation to their Medicare and Medicaid legislation requiring that certain elements be included on medical records. The medical insurance industry is demanding more complete documentation for reimbursement. The Joint Commission on Accreditation of Hospitals (JCAH) and PSRO groups are also putting demands on medical record keeping for purposes of quality control. In the area of forensic medical events such as rape, child abuse, assault and others, proper record keeping is important from a legal standpoint. In plaintiff litigation the quality of the medical record is extremely important. In fact, it is crucial for a successful defense against a medical malpractice suit.

Medical records have three functions. The first and most important, is that of medical documentation. Good medical record keeping is not an obsessive-compulsive personality derangement. Good medical record keeping is good medicine, plain and simple. The basic elements of medical record keeping are those of disease evaluation, treatment, progress, and disposition. The second function of the medical record is that of communication. This may be between a physician and patient, or physician and physician. The third function is that of legal documentation. The best legal document is a good medical document.

There is really no universal medical record form. A form must be tailored to suit its individual setting. Forms suitable for a private community hospital which has a fixed physician staff might not be suitable to a teaching hospital where the staff changes frequently. The latter has to be more structured and specific in the components it includes. Despite such individualization, there are general principles that can be applied to all medical records. The record must be direct, complete, and as simple as possible. Whatever components are to be included in the medical record should be so stated. It should be as time-saving as possible. An area which can be designed to save time is one listing laboratory tests as a check-off list rather than a write-in area.

In designing a medical record, one should be aware that no record is permanent. It should be designed on an experimental basis. Most institutions do not order forms to last a year, two years or five years, but rather in specific small quantities on a 3 month or 6 month basis. Changing a medical record usually does not add to its cost. Many printing firms have a standard charge for a quantity of forms and no additional charge for development or changes. Therefore, if a particular chart does not work well, it can be altered inexpensively. Communication with the printing vendor regarding his mechanical limitatons is essential in order to determine his ability to produce a chart of desired specifications.

There are three basic medical record design types. The first — and probably the most widely used — is a single page form with multiple cabon copies. These carbon copies have various ultimate destinations — one may remain in the emergency room while the rest are dispatched to medical records, the billing department, and other areas. This has the advantage of avoiding the need for photocopying of the record for distribution. The greatest disadvantage is that such forms necessarily have limited space for medical documentation.

The second type consists of multiple pages with no duplicate copies. This type of form provides much more recording space, but requires photocopying. The third kind has multiple pages with some carbons allowing for duplication of parts of the record. The last two forms described are more expensive because they are more intricate and require more effort on the part of the printer.

The standard format of medical records is one sequentially listing history, physical examination, diagnosis, and treatment. This is a very static format and it does not lend itself well to chronological documentation. There are those who favor the problem-oriented format, which allows for a more dynamic chronologic entry of medical events. Such a record is a virtually blank sheet with columns on one side for time listing and another column for the entry, whether this be an objective examination, subjective information, plan or treatment. This form lends itself to entries made by multiple parties. Such parties might include the treating physician, consultants, nurses, triage officers, respiratory therapists and others caring for a patient. In principle this seems like a very good medical record, but in fact it is quite difficult to institute.

Figure 1 is an example of a single-page multiple-copy type of medical record. The major disadvantage is obvious. The majority of the sheet is devoted to administrative documentation and little area is left for physicians' notes. Figures 2 through 6 show the record we developed at the University of Chicago. This is a multiple-page chart with no duplicates. The top portion of each of the pages contains patient identification areas. The first page (Fig. 2) also contains the time of admission, vital signs, triage notes, a listing of diagnostic tests, medication and treatment administered, and instructions to the patient. Page two (Fig. 3) consists of a half sheet which duplicates the diagnostic test area of the first page. This is a clerical tear-out. The physician ordering laboratory tests indicates the desired tests on the first page and gives the clerk the carbon copy. The clerk in turn uses this copy to complete the appropriate requisition forms. The third page (Fig. 4) is the patient's copy. It allows for duplication of the patient information found at the bottom of the first page. It also contains various pre-printed patient instructions for a number of entities — wound care, head injury, and others — the appropriate portions of which can be checked. The fourth page (Fig. 5) is a largely blank sheet for the physician to record his history, physical findings and diagnosis. A good deal of space is allowed for this

EMERGENCY ROOM REPORT

FORKOSH MEMORIAL HOSPITAL
2544 W. MONTROSE AVE., CHICAGO, ILL. 60618

REPORT NO. 465916-5

NAME (LAST) (FIRST) | HOME PHONE | DATE | TIME A.M. P.M.
SOCIAL SECURITY NO. | BIRTH DATE | AGE | SEX M F
ADDRESS | CITY | STATE | ZIP | CIVIL STATUS S M W D SEP
NEAREST RELATIVE OR FRIEND (IF MINOR, PARENT'S FULL NAME) | RELATIONSHIP | RELIGION
ADDRESS | CITY | STATE | PHONE | LAST RITES BY
EMPLOYER | PHONE | RACE
ADDRESS | CITY | STATE | OCCUPATION
BLUE CROSS NO. | CODE NO. | EFFECTIVE DATE | ☐ SUBSCRIBER ☐ DEPENDENT | COMP. CASE
GROUP & SUBSCRIBER NO. | SUBSCRIBER'S NAME
PUBLIC AID CARD NO. | EXPIRATION DATE | CASE NAME ON CARD
MEDICARE CARD NO. (INCLUDE ALL LETTERS & NOS.)

E.R. PERSONNEL PLEASE OBTAIN AS MUCH INSURANCE INFORMATION AS POSSIBLE

BROUGHT BY: SELF POLICE FIRE RELATIVE OTHER | FIRE AMBUL. NO.
OFFICER'S SIGNATURE | STAR NO. | DISTRICT
DESCRIPTION OF ACCIDENT OR INJURY - STATE HOW, WHEN AND WHERE OCCURRED - GIVE NAME OF PERSON REPORTING

NOTIFIED
RELATIVE ______ NAME
POLICE ______ NAME
CORONER ______ NAME
PRIVATE DOCTOR
DOCTOR'S PHONE NO.
BY WHOM
TIME A.M. P.M.
ALLERGIES? ______
EPILEPTIC? ______
DIABETIC? ______

PHYSICIAN'S REPORT | SIGNATURE OF NURSE

CONDITION ON ADMISSION: GOOD ☐ FAIR ☐ POOR ☐ HEMORRHAGE ☐ COMA ☐ TEMP. ORAL RECTAL P. R. B.P. /

STATE WHAT X-RAYS WERE TAKEN | READING: POSITIVE ☐ NEGATIVE ☐

DIAGNOSIS:

TREATMENT:

HYPER-TET UNITS | TETANUS TOXOID | AMOUNT | GIVEN BY | TIME A.M. P.M.
DRUGS GIVEN | ANESTHESIA
SOLUTIONS | NO. AND TYPES OF SUTURES
CONDITION ON DISCHARGE ☐ GOOD ☐ FAIR ☐ OTHER DESCRIBE | NAME OF STAFF DOCTOR NOTIFIED | BY | TIME A.M. P.M.

DISPOSITION OF CASE | ATTENDING DOCTOR'S SIGNATURE

☐ HOME
☐ HOME - ADVISED TO SEE FAMILY DOCTOR
☐ HOME - ADVISED TO SEE DOCTOR ON CALL
☐ HOME - RETURN IN A.M.
☐ ADVISED TO GO TO ______ HOSPITAL
☐ ADMITTED TO ROOM ______
☐ SIGNED RELEASE - WENT HOME
☐ SIGNED RELEASE - WENT TO ______ HOSPITAL
☐ TRANSFERRED TO: ______
VIA ☐ POLICE ☐ FIRE ☐ FAMILY ☐ PRIVATE AMBULANCE ☐ OTHER

AUTHORIZATION FOR TREATMENT, OPERATION AND RELEASE OF INFORMATION

I hereby consent to and authorize the giving of all treatments, the performance of all operations, the administration of all anesthetics and any and all other technical procedures which, in the judgement of the attending physician, may be considered necessary or advisable for the diagnosis or treatment of this case while being treated as an outpatient in Forkosh Memorial Hospital. Further, I hereby authorize said Forkosh Memorial Hospital to release any medical information required by any insurance company or other third party in connection with the indemnification or payment by it of any portion of the related hospital bill.

WITNESS ______ | (SIGNATURE OF PATIENT, NEAREST RELATIVE, OR LEGAL GUARDIAN)
SIGNED FOR PATIENT BY ______ | RELATIONSHIP | DATE | TIME A.M. P.M.

WAIVER OF RESPONSIBILITY FOR DISCHARGE

______ (NAME)
IS LEAVING THE HOSPITAL AGAINST THE ADVICE OF THE ATTENDING PHYSICIAN AND AGREES NOT TO HOLD THE DOCTOR AND THE FORKOSH MEMORIAL HOSPITAL RESPONSIBLE FOR ANY HARM OR INJURY THAT MAY RESULT FROM MY ACTION.

WITNESS | SIGNED

ASSIGNMENTS — HOSPITAL AND DOCTOR AUTHORIZATION TO PAY INSURANCE BENEFITS

I hereby authorize payment directly to the Hospital and/or Doctors named above of the applicable insurance policy benefits otherwise payable to me, but not to exceed their charges. I understand I am financially responsible to them for charges not covered by this assignment.

PATIENT ______ OR ______ (RELATIVE) (RELATIONSHIP TO PATIENT)

MEDICAL RECORD COPY

Figure 1.

DATE: TIME IN			TRIAGE: SEEN IN		AGE:
			☐ A · AMBULATORY CLINIC		SEX:
E.R. TRANSPORT	NAME & NO.		☐ E · EMERGENCY SECTION		RACE:
VITAL SIGNS	B.P.	TEMP.	☐ X · E.R. WHEN AMB. CLOSED		LOG NO.
	PULSE	RESP.	☐ D · DENTAL CLINIC		TRIAGE OFFICER
			☐ W · GYNE CLINIC		CLERK

IS THIS PATIENT CONNECTED WITH THE UNIVERSITY OF CHICAGO? YES ☐ NO ☐ NAME PHONE

INJURY COMPLAINT AND TRIAGE NOTES
ALLERGIES — LAST TETANUS
PRESENT MEDS

DIAGNOSTIC TESTS

☐ GLUCOSE	☐ OTHER				☐ U/A C.C.		☐ U/A CATH		☐ URINE C&S			
☐ ELECTROLYTES (NA	CL	K	CO₂)	☐ PG	MACRO							
☐ POTASSIUM	☐ BARBITURATE	PT			MICRO							
☐ CALCIUM	☐ SALICYLATE	PTT			☐ ECG							
☐ BUN	☐ ALCOHOL	CBC	WBC									
☐ CREATINE	☐ FIBRINOGEN	DIFF	RBC									
☐ AMYLASE	☐ MAGNESIUM	META	Hgb	☐ X-RAY								
☐ TOT. BILI	☐ LACTATE	BAND	Hct									
☐ CPK	☐ ACETONE	SEG	MCV	☐ CULTURE		TYPE						
☐ LDH	☐ OSMOLALITY	LYMPH	MCH				☐ CO₂					
☐ SGOT	☐ CSF. PROTEIN	MONO	MCHC	ABG #1 AT:	FIO₂	pH.	pCO₂	pO₂				
☐ SGPT	☐ CSF. GLUCOSE	EOS	☐ RETIC	ABG #2 AT:	FIO₂	pH.	pCO₂	pO₂				
☐ ALK. PHOS	☐ UR. AMYLASE	BASO	SED RATE	T & C	units of							
☐ HBDH	☐ UR. OSMOLALITY	ATYLYMPH	PLATLET	☐ OLD MEDICAL RECORDS								
☐ OTHER		MORPH	☐ SICKLE SCREEN	☐ ED. RECORDS FROM								

DOCTOR'S ORDERS

TIME	DOCTOR	MEDICATIONS AND TREATMENT	TIME	R.N.

DIAGNOSIS

DISPOSITION

☐ HOME ☐ ADMIT ☐ LWBS ☐ AMA ☐ TRANSFER   WHERE _____ HOW _____   APPROVED BY

☐ SERVICE _____   TIME OUT _____   ACCEPTED BY

INSTRUCTIONS TO PATIENT

FOLLOW UP: ☐ RECHECK: RETURN TO BILLINGS EMERGENCY ROOM BETWEEN 8-10 A.M. ON _____ SEE DR _____
☐ CLINIC REFERRAL: CLINIC NAME _____ (SEE PINK REFERRAL SHEET)
☐ PRIVATE PHYSICIAN OR COMMUNITY CLINICS

☐ MEDICATIONS & DOSAGES: _____ (FOLLOW LABEL INSTRUCTIONS FOR PRESCRIPTION FROM THE EMERGENCY ROOM)

☐ MEDICALLY AUTHORIZED TIME OFF FROM WORK/SCHOOL UNTIL _____

☐ AFTERCARE INSTRUCTIONS: ☐ WOUND CARE ☐ HEAD INJURY ☐ PELVIC INFECTION ☐ VIRAL INFECTION
☐ THREATENED ABORTION ☐ SPRAINS FRACTURES & BRUISES ☐ ABCESS ☐ LOW BACK STRAIN

☐ OTHER _____

_____ R.N.
_____ M.D.

PATIENT'S SIGNATURE: I HEREBY ACKNOWLEDGE THE RECEIPT OF THE ABOVE INSTRUCTIONS.

CHART - 1

Figure 2.

DATE: TIME IN

TRIAGE: SEEN IN

A -	AMBULATORY CLINIC
E -	EMERGENCY SECTION
X -	E R WHEN AMB CLOSED
D -	DENTAL CLINIC
W -	GYNE CLINIC

AGE:

SEX:

RACE:

LOG NO.

TRIAGE OFFICER

CLERK

E. R. TRANSPORT   NAME & NO.

NAME

PHONE

VITAL SIGNS

B.P.	TEMP.
PULSE	RESP.

IS THIS PATIENT CONNECTED WITH THE UNIVERSITY OF CHICAGO?   YES ☐   NO ☐

LAST TETANUS

INJURY COMPLAINT AND TRIAGE NOTES

ALLERGIES

PRESENT MEDS

DIAGNOSTIC TESTS

☐ GLUCOSE	☐ OTHER	☐ U/A C.C.	☐ U/A CATH	☐ URINE C&S
☐ ELECTROLYTES (NA | CL | K | CO₂ |)	☐ PG	MACRO:		
☐ POTASSIUM	☐ BARBITURATE	☐ PT	/		MICRO:		
☐ CALCIUM	☐ SALICYLATE	☐ PTT	/		☐ ECG:		
☐ BUN	☐ ALCOHOL	☐ CBC	☐ WBC			
☐ CREATINE	☐ FIBRINOGEN	☐ DIFF	☐ RBC			
☐ AMYLASE	☐ MAGNESIUM	☐ META	☐ Hgb		☐ X-RAY:	
☐ TOT. BILI.	☐ LACTATE	☐ BAND	☐ Hct			
☐ CPK	☐ ACETONE	☐ SEG	☐ MCV			
☐ LDH	☐ OSMOLALITY	☐ LYMPH	☐ MCH		☐ CULTURE	
☐ SGOT	☐ CSF PROTEIN	☐ MONO	☐ MCHO		☐ ABG #1 AT:	FIO₂:	pH:	CO:
☐ SGPT	☐ CSF GLUCOSE	☐ EOS	☐ RETIC		☐ ABG #2 AT:	FIO₂:	pH:	pCO₂:	pO₂:
☐ ALK PHOS	☐ UR AMYLASE	☐ BASO	☐ SED RATE		☐ T & C	units of		pCO₂:	pO₂:
☐ HBDH	☐ UR OSMOLALITY	☐ ATYLYMPH	☐ PLATLET		☐ OLD MEDICAL RECORDS	
☐ OTHER		☐ MORPH	☐ SICKLE SCREEN		☐ ED RECORDS FROM	

TYPE

CLERK - 2

Figure 3.

Michael C. Tomlanovich, M.D.

Figure 4.

78

DATE: TIME IN			TRIAGE: SEEN IN		AGE:	
			☐ A - AMBULATORY CLINIC			
					SEX:	
E.R. TRANSPORT	NAME & NO.		☐ E - EMERGENCY SECTION		RACE:	
VITAL SIGNS	B.P.	TEMP.	☐ X - E.R. WHEN AMB CLOSED		LOG NO.	
	PULSE	RESP.	☐ D - DENTAL CLINIC		TRIAGE OFFICER	
			☐ W - GYNE CLINIC		CLERK	
IS THIS PATIENT CONNECTED WITH THE UNIVERSITY OF CHICAGO? YES ☐ NO ☐			NAME			PHONE

INJURY COMPLAINT AND TRIAGE NOTES	ALLERGIES	LAST TETANUS
	PRESENT MEDS	

HISTORY FINDINGS DIAGNOSIS

_____ M.D.

(IF NEEDED, CONTINUE ON OTHER SIDE)

CHART - 4

Figure 5.

DATE TIME IN			TRIAGE SEEN IN	AGE		
			[ ] A - AMBULATORY CLINIC	SEX		
E.R. TRANSPORT	NAME & NO		[ ] E - EMERGENCY SECTION	RACE		
VITAL SIGNS	B.P.	TEMP	[ ] X - E.R. WHEN AMB CLOSED	LOG NO		
			[ ] D - DENTAL CLINIC	TRIAGE OFFICER		
	PULSE	RESP	[ ] W - GYNE CLINIC	CLERK		
SOCIAL SERVICE			PATIENT'S NAME (PRINT ONLY)			PHONE

MODE OF ARRIVAL [ ] AMBULATORY [ ] STRETCHER [ ] WHEEL CHAIR

CLOTHING & VALUABLES

STATUS ON ARRIVAL IN EMERGENCY ROOM BP P R TEMP [ ] ORAL [ ] RECTAL

[ ] TO FAMILY

APPEARANCE

[ ] ALERT ORIENTED RESPONSIVE   [ ] DISORIENTED BUT RESPONSIVE

[ ] LOCKED UP   [ ] COMATOSE [ ] SOMNOLENT [ ] OTHER (SKIN COLOR ETC.)

[ ] WITH PATIENT   CHIEF COMPLAINT

DATE	HOUR	UNUSUAL PROBLEMS	EXPECTED OUTCOME	SIGNATURE

DATE	HOUR	NURSING ORDERS	SIGNATURE

DATE	HOUR	CLINICAL NURSING NOTES	SIGNATURE

NURSING - 5

Figure 6.

medical documentation. The vital signs are duplicated from the front sheet via carbon copy. The fifth page (Fig. 6) is for nursing notes.

A major disadvantage of such a record is the absence of duplicate copies. Copies which are needed must be made by photocopying of the desired portions of the record. This record lends itself well to documenting the course of truly emergent patients — those who require extensive documentation of procedures performed upon them and their course in the Emergency Department. An Emergency Department dealing with a low volume of critically ill patients may not require a medical record of this degree of sophistication.

Figures 7 through 9 depict a multiple-page record which includes some duplicates. The first page (Fig. 7) contains a small section for administrative details in its upper portion. The remainder of the page is devoted to the physician's observations, medical procedures, and tests. There are two duplicates to this page. The first duplicate is kept in the Emergency Department, the second is sent to the billing department. The fourth page of the record (Fig. 8) is a billing page. The clerical people, who must, from their copy of the medical record, calculate what charges the patient has incurred, use this page to itemize charges. The fifth page (Fig. 9) is a duplicate of the first three in its lower portion. The space above this is devoted to written instructions to the patient. This is a legible transcription by the clerical staff of the physician's orders and information to the patient.

The Joint Commission on Accreditation of Hospitals has formulated specific requirements for the content of emergency department medical records. The requirements include patient identification, the components of which are name, address, telephone

Henry Ford Hospital   **DIV. OF EMERGENCY MEDICINE**   **NO. E 20501**   DATE-TIME STAMP

PATIENT'S LAST NAME   FIRST   MIDDLE   MAIDEN

MRN

ADDRESS   CITY   STATE   ZIP   PHONE

BIRTHDATE   AGE   ☐ MALE   BILLING CODE   BROUGHT TO D E M BY   EMS _____   ALLERGIES? ☐ NO ☐ YES (SPECIFY)
☐ FEMALE   AMB _____ OTHER _____

CURRENT MEDICATION   ☐ NO ☐ YES (SPECIFY)   DIABETIC ☐ NO ☐ YES

PATIENT COMPLAINT:

TRIAGE

TIME   TRIAGE CAT   T   BP   P   R
☐ I   LACERATION LOCATION-LENGTH
☐ II   IN
☐ III   IN

PROBLEM:

SUBJECTIVE: | **MEDS/IV'S**

OBJECTIVE:

PROCEDURES DONE

DIAGNOSIS | ☐ CONSULT DR _____ (SERVICE)

PRESCRIPTIONS | **LAB. HICDA**

TESTS

☐ Hgb & Hct   ☐ Hgb & WBC   ☐ HEMA A   ☐ HEMA B
☐ U/A _____   ☐ CREAT _____   ☐ SMEAR ☐ CULTURE OF
☐ B SUGAR _____   ☐ AMYLASE _____
☐ BUN _____   ☐ OTHER _____   FOR _____ RESULT _____
☐ OTHER _____   ☐ LYTES PROFILE _____   EKG RESULT _____
☐ OTHER _____

X-RAYS

☐ ABDOM   ☐ CHEST
☐ SKULL
☐ EXTREM   RESULT
☐ OTHER   RESULT
☐ OTHER   RESULT

NOTIFICATIONS: ☐ POLICE ☐ MED. EXAMINER ☐ BD. OF HEALTH ☐ SOCIAL SERV.
☐ VNA ☐ OTHER _____

DISPOSITION: PATIENT ☐ HOME ☐ TRANSFER TO _____
☐ ADMIT ☐ OTHER _____
VALUABLES ☐ NONE ☐ PATIENT ☐ HOME ☐ SECURITY ☐ CASHIER
CLOTHES ☐ NONE ☐ PATIENT ☐ HOME ☐ SECURITY ☐ PROPERTY RM

NURSE'S NAME (PRINT)
NURSE'S SIGNATURE   TIME
PHYSICIAN'S NAME (PRINT)
PHYSICIAN'S SIGNATURE

FOR PATIENT EDUCATION:

RELE

PRECAUTION SHEETS GIVEN: ☐ WOUND ☐ HEAD ☐ CAST ☐ VD ☐ SPRAINS ☐ DOG BITE ☐ NOSE BLEED ☐ OTHERS
☐ PEDS   DISCHARGE TIME _____

MEDICAL RECORDS DEPT.

FORM 603620 12-76

Figure 7.

Henry Ford Hospital   **DIV. OF EMERGENCY MEDICINE   NO. E 20501**

DATE-TIME STAMP

PATIENT'S LAST NAME	FIRST	MIDDLE	MAIDEN

MRN

ADDRESS	CITY	STATE	ZIP	PHONE

BIRTHDATE	AGE	☐ MALE ☐ FEMALE	BILLING CODE	BROUGHT TO D.E.M. BY:	EMS _____ AMB _____ OTHER _____

☐ OPD 20   ☐ IPD 25   DOC.   CODE   HICDA

SERVICE DESCRIPTION	32	SERVICE CODE	37	38	CHARGE	44
Emergency Treatment Room	9	0 9 6 0 1				
Emergency Room Facility	9	0 9 6 0 2				
Anesthesia - Local	9	1 0 0 0 0				
Injection(s)	0	8 9 3 0 0				
Medical Emergency	9	2 0 3 6 8				
Emergency First Aid	9	2 0 3 6 9				
Exam	0	1 0 0 0 0				
Tray	7	1 5 0 0 0				
ABG	0	2 2 3 5 1				
EKG	0	6 8 9 5 7				
	TOTAL CHARGE					

ACCOUNTING

Figure 8.

83

NOTIFICATIONS: ☐ POLICE ☐ MED. EXAMINER ☐ BD. OF HEALTH ☐ SOCIAL SERV.
☐ VNA ☐ OTHER _____

DISPOSITION: PATIENT ☐ HOME ☐ TRANSFER TO _____
☐ ADMIT ☐ OTHER _____

VALUABLES ☐ NONE ☐ PATIENT ☐ HOME ☐ SECURITY ☐ CASHIER

CLOTHES ☐ NONE ☐ PATIENT ☐ HOME ☐ SECURITY ☐ PROPERTY RM.

FOR PATIENT EDUCATION:

PRECAUTION SHEETS GIVEN: ☐ WOUND ☐ HEAD ☐ CAST ☐ VD ☐ SPRAINS ☐ DOG BITE ☐ NOSE BLEED ☐ OTHERS
☐ PEDS

NURSE'S NAME (PRINT) _____

NURSE'S SIGNATURE _____ TIME: _____

PHYSICIAN'S NAME (PRINT) _____

PHYSICIAN'S SIGNATURE _____

DISCHARGE TIME _____

RELEASE

PATIENT

FORM 603620 12-76

84

Figure 9.

number, age, sex, and date of birth. There must be a description of the patient's arrival at the hospital, which includes the time as well as the mode of transportation. Other requirements are the standard ones of history, physical examination, findings, and disposition. The results of diagnostic tests must appear on the medical record. Recording such results is very useful, especially if extensive patient follow-up is conducted in the Emergency Department itself. Additionally, details of treatment and the patient's condition at the time of disposition are required. Many medical records do not include these items of information, but they are very important. The J.C.A.H. also requires a patient log to be kept as a separate document from the medical record. The log must contain the name, date, diagnosis, treatment, and disposition of the patient.

There is a considerable number of items whose inclusion on emergency medical records is essential. If record space is appropriately budgeted, these items can all be included on a medical form of appropriate size. The first of these is a record of times. This includes the time the patient was registered, the time he was first seen by any medical person, when he was evaluated by a physician, and his time of departure from the area. Such times are typically documented in various places about the record — usually adjacent to the signatures of physicians, nurses, clerks, and triage personnel. A record of these times becomes very useful if an assessment of the department is being made. If problems of patient flow are being encountered, medical records can be audited. An assessment can be made as to which portion of patient care is consuming the longest time by examination of these entries. Additionally, these times are useful in determining patient progress.

Components of biographical data and insurance information may expand *ad nauseum*. Many emergency records include not only the essential biographical data, but also race, religion, occupation, name of employer, closest living relative and other information. Such items, when combined with insurance information — contract number, policy number, data of effectiveness, date of expiration, — become so space-consuming that they occupy space that would be better utilized for medical documentation. These components are important for billing and administration purposes, but not in medical record function. Administrative information areas consume a majority of the space of most single-page medical records. Such information should ideally be put on a totally separate form, or, in hospitals having a computer terminal in the emergency department — as we do at Henry Ford Hospital — fed directly into a computer. In this way, the information does not have to be put on paper at all. Another advantage to computerizing such information is rapid access. A read-out of administrative data can be instantly retrieved by merely entering a patient's medical record number.

Another very important component of a medical record is that portion which allows for patient consent. When a multiple-page record is used, this is best placed on the back of the first page, because this page usually becomes a part of the patient's permanent medical record. Consent does not have to be duplicated on each page. Consent forms should be complete and include whatever administrative and medical functions a hospital requires. The components should be drawn up with advice of a knowledgeable legal counsel. An essential component is authorization for treatment. This may be worded in a number of ways. There are medical records which authorize any examination, medical treatment, or surgical treatment, under local or general anesthesia deemed necessary. I doubt that legality such a consent would bear in a court of law. Dr. Segal has referred to the limitations of written consent in the Emergency setting (see chapter entitled Medical-Legal Policies, Procedures, Liability and Consent — ed.) but he utilizes written consent forms in his own emergency department, so he must acknowledge to some extent their usefulness as legal documents.

Additional clauses may be put into the consent form to the effect that the patient acknowledges the accuracy of the information he has been given, or that he authorizes release of information to appropriate insurance carriers or employers. The latter is a provision in workman's compensation cases. A law recently passed in Michigan bans the release by a hospital or physician of information regarding alcohol and drug abuse without a *specific* written consent. Under this law, authorization for release of such information on a general release form is not adequate.

Another component of the consent form consists of the acceptance of financial responsibility by the patient. Photographic permission is a component often included at academic institutions. Permission is requested to take, retain, and publish photographs for educational and scientific purposes. Hospitals actively involved in medical education might want to have this as part of their consent. Another component sometimes included is a statement to the effect that the patient understands the nature of emergency care, that he realizes that this care is not necessarily complete, and that he may need to seek additional care.

Vital signs are an obviously important part of the medical record. Triage notes are a very important component as well. They are important regardless of whether or not a decision to send the patient to another area of a hospital is made on their basis. Often the triage note will constitute the written record of the first medical individual to have contact with the patient. Triage notes include the chief complaint, current medications, history of allergies and immunizations, and other medical history. A well-written triage note listing this information can save the physician writing time in documenting such history.

Triage or nurses' notes do not, however, substitute for a physician's history. Physician history, therefore, becomes another necessary component of the medical record. Physical examination is another component, as are laboratory tests. There is some controversy over whether laboratory tests should be listed on the record in a check-off format or whether all such tests should be written in individually. The check-off system is time-saving, but those who oppose it maintain that it leads to the over-ordering of laboratory tests by physicians. Some medical records consolidate laboratory tests and x-rays, while others place these two entities in separate areas. Additionally, there must be an area which allows for documentation of the fact that a consultant was called as part of the treatment of a patient. In the area of treatment, both procedures and medications must be documented.

The diagnosis and disposition are also important parts of the medical record. Problem-oriented records include both an intial assessment and final diagnosis. We utilized this form of chart, but discontinued it because we found that the records often read very poorly. For examply, the initial impression of a house officer might be "acute myocardial infarction", but this diagnosis disproven on further evaluation, and the patient sent home. Such a record might, nonetheless, indicate that we are sending home patients with acute myocardial infarction.

Notification is an area which can be included on the medical record. Many states require the notification of various agencies about certain medical events. In Michigan this includes notification of the police of all criminal assault, rape, and child abuse. Hospitals are also required by law to notify the medical examiner of deaths on arrival. The Board of Health requires notification regarding various infectious diseases.

Nurses' notes are an optimal component of the main medical record. Not every case requires nurses' notes. This is very much dependent on the nature of the case. A patient with a twisted ankle hardly requires nurses' notes, but the very critically-ill patient does, and such notes constitute an extremely important component of the medical record in these cases.

Disposition is another important component. This should include whether the patient was sent home, admitted or transferred, the time that this occurred, whether any prescriptions were written for him, what the physician's instructions were, and what follow-up was agreed upon.

The recording of the patient's condition on discharge on the medical records is a J.C.A.H. requirement. This is a very important item of information, but one that is frequently not included in records. This information is especially useful in cases involving patients whose conditions change remarkably in response to treatment administered in the emergency room. This includes hypoglycemia, asthma, and other life-threatening entities for which patients can be given appropriate emergency treatment and eventually discharged.

In addition to standard emergency department records there is a need for a number of special forms.

A facility which cares for a large number of critically-ill patients will have need for a critical care flow sheet. The standard medical record is not spacious enough to allow for adequate documentation in such cases, and is also typically a static document which does not allow for ongoing chronological documentation of a patient's treatment and progress. Critical care flow sheets can serve this function.

Sexual assault and child abuse reports often need to be filed on special forms. Such forms should be designed in cooperation with medical-legal and law enforcement experts. Ideally, they should be standardized state-wide to insure uniformity in medical documentation and medical evidence-gathering. Many states require the reporting of certain disease entities. Therefore, special forms may be needed for this purpose, although they will only rarely be used, because there are very few reportable entities for which the definitive diagnosis can be made in the emergency department. Some hospitals utilize patient transfer forms. Unfortunately, the completion of such forms provides additional writing for the physician or nurse. It is our policy to merely send a photocopy of the medical record along with the patient when we transfer him to another institution.

Patient isntruction sheets are another type of highly desirable special form. Our institution has developed separate sheets for instructions regarding various conditions and disease entities. Other institutions have placed multiple-entity instructions on a single sheet of paper. This has the obvious advantage of conserving paper, but other than this there are no advantages to this format. The information on such forms does not substitute for the carefully-given instructions of a physician to his patient. During litigation, patients have claimed that instruction forms were never explained to them, and physicians' defenses based upon written instruction given to patients have not held up well. Each of our forms, nevertheless, provides space at the bottom for a patient's signature. Instruction forms are required by the J.C.A.H. We spend a great deal of time on our instruction forms in order to make them as clear as possible. We have tried to make them complete as well as comprehensible. Some areas of instruction covered by such forms are wound and suture care, sprains, venereal disease, head injury, nose bleed, and cast care. A hospital which sees a high volume of pediatric patients may need to develop a separate set of pediatric instruction forms detailing treatment of fever, diarrhea, and other entities.

A well-organized patient log will provide necessary information regarding patient visits in short order. This is especially important to a department performing any extent of patient follow-up. In addition, it is wise to retain copies of recent emergency department records in the area for a short time. At our institution, we maintain an active file for thirty days. We then keep records in a less accessible back room for an additional six months and discard them after that. The original copy of course, goes to the permanent medical record, where it remains indefinitely.

CHAPTER 8

EMERGENCY DEPARTMENT ARCHITECTURAL DESIGN

Lawrence R. Drury, M.D.

If the emergency physician is fortunate enough to be consulted when plans are being made to rebuild or remodel his facility, he should contemplate what occurs in the Emergency Department and what functions are performed there. That he takes care of ill and injured patients is obvious. He must ask himself how physical structure effects his ability to do this. An even more important issue is how he can utilize the design of the Emergency Department to better process and care for such patients. Only by examining the functions of the Emergency Department can one develop concrete suggestions regarding how to improve it.

My remarks about Emergency Department design are aimed mainly at the community hospital Emergency Department that has ten to fifty thousand patient visits per year. Though some of the information is applicable to larger urban institutions, it is generally directed to smaller hospitals. Relevant issues are those of the actual use of space in the Emergency Department, the division of space by function, calculation of space needs, and the factors that determine a decision to remodel or rebuild.

Two functions take place with regard to patients in the Emergency Department: The actual treatment of patients and waiting or holding. The latter occupies a good deal of time. Patients wait after being registered, they wait for laboratory work to be drawn, and those results to return. They then wait again for the physician to confer with them regarding these results. If they require admission or consultation by other physicians, this requires more waiting time still. When one analyzes these functions on the basis of the time they consume, it becomes apparent that waiting occupies much of the patient's time. In a patient visit lasting forty-five minutes, for example, probably no more than fifteen involve contact with medical personnel, with the remainder of the time spent waiting. An emergency department design that would best take this into account would be one with a small well-equipped and staffed treatment space and a larger, less sophisticated waiting area.

I have seen no emergency departments that are actually built along these lines. Many display the traditional design, an example of which is seen in Figure 1. This design includes various areas which are devoted to the traditional specialties: pediatrics, medicine, surgery, psychiatry and gynecology. These are, in turn, divided into small cubicles, Each cubicle is intensely built, in that each contains a relatively large amount of equipment and has a large staffing requirement. Since patients spend much of their time waiting, such a setting is wasteful of space, equipment, and personnel.

The design that best meets the problems of a modern emergency department is that of a central nursing desk surrounded by examination cubicles, a

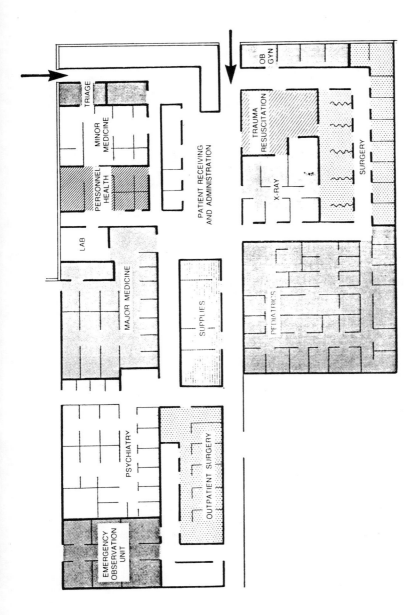

Figure 1.

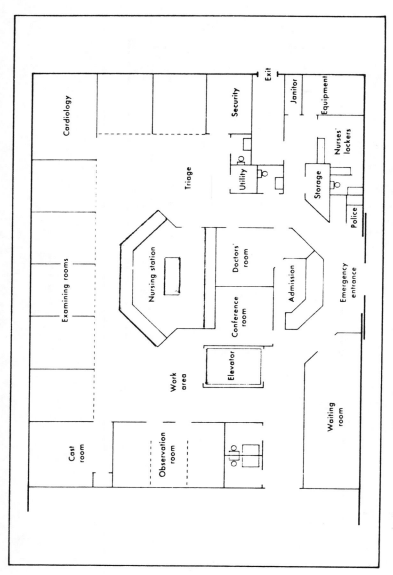

Fig. 1-2. Emergency department: arena-type floor plan.

Figure 2.

design pioneered in the intensive care unit. This is the most convenient type of area for nurses and physicians, because working in it requires less walking. Patient care is improved because patients can be observed closely. This design is quite adaptable to change should this become necessary. It is probably the least expensive to build and to modify. Figure 2 shows an example of this design. There are twelve central examining cubicles, and an area devoted to cardiology. There is no specific trauma resuscitation area, so the central section would have to be used for this. Figure 3 shows a similar plan, but one which includes more specialty areas.

Figure 4 depicts yet another plan. The patient entrance is at the bottom. Patients enter the registration area, are triaged by a nurse to one of the sixteen examination beds. In addition to these beds, there is a cast room and an ear, nose and throat room. There is also an area where sigmoidoscopy can be performed. This can double as a gynecological examination room, because it has a wall surrounding it and therefore offers the necessary privacy. The door on the right side leads to the intensive care unit. The main x-ray department is across the hall on the other side. There is also a two room x-ray facility within the emergency department itself. This design is utilized by an Emergency Department in the Chicago area which has about 40,000 patient visits per year. This emergency department also has a unique supply system. Supplies are kept in a bank of drawers against each wall. The supply banks are all identical to each other.

There are no walls separating the cubicles numbered one through sixteen. The dotted lines represent curtains. Therefore, due to the uniform supply arrangements, beds may be utilized for all types of patients. One would expect patients to complain about

Lawrence R. Drury, M.D.

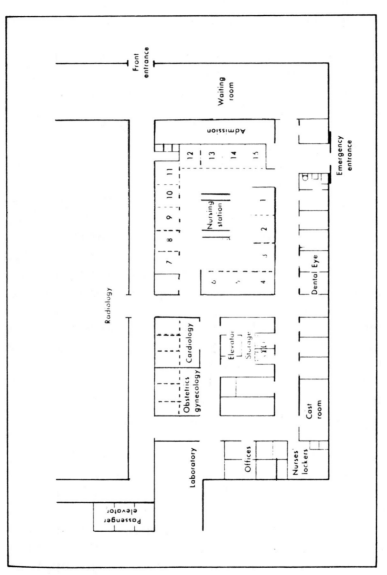

Emergency department: central work core (three-corridor floor plan).

Figure 3.

98

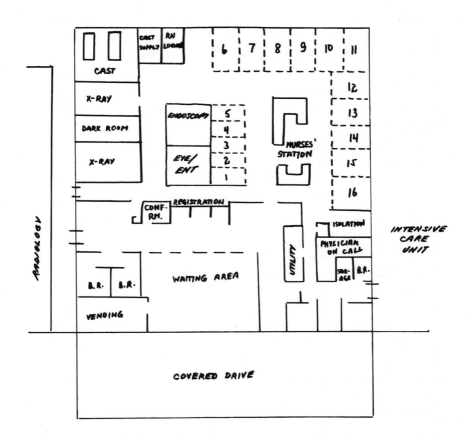

Figure 4.

99

the lack of privacy in such a set-up. However, this has so far not been the case in this facility, which has been functioning for six or seven years.

Figure 5 represents a planning blueprint of a university hospital emergency department. There are two entrances into the area. Security is provided by a security guard stationed within a glass-enclosed room with visibility over the entire department. The main body of the treatment area is broken up by walls into discrete areas. These areas can be dedicated to specific functions if this is desired. Visual contract with the main treatment area is afforded from the central nursing desk. Administrative areas, offices and classrooms are also included. The two dark areas represent seating areas, one for relatives and one for ambulatory patients waiting for laboratory and x-ray results.

How does one determine the number of seats needed in a waiting area? The average patient comes to the emergency department with one friend, relative or other concerned person. Therefore, knowing the maximum number of registrants per hour as well as the average turn-around time, one can calculate the seats necessary to accommodate the people accompanying patients to the hospital. For example, an emergency department caring for a maximum of twenty patients per hour with a turn-around time of an hour will require twenty seats. If the turn-around time is 3 hours, this figure increases to 60 seats. Failure to provide such a space will result in relatives leaving the area. This can be a problem as they are often instrumental in providing crucial patient history. The seating area should be made humane. There should be access to bathrooms, and there might be some form of entertainment provided. Television is a good device to take visitors' minds off their waiting. There ought to be vending machines, and a number of public tele-

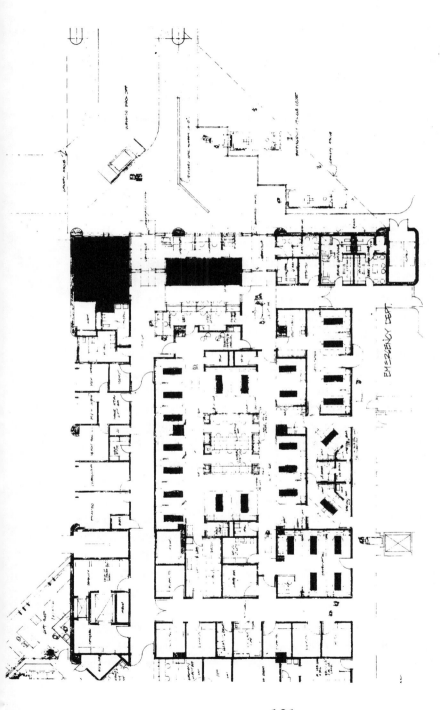

Figure 5.

phones as well. For the hospital with a significant volume of pediatric patients it might be useful to create a separate pediatric waiting area with toys and other play equipment. A waiting area should be far enough away from the registration and triage areas to provide privacy for the patients going through these areas.

I am a strong advocate of triage areas for all hospitals. The first person that a patient encounters should be one who is charged with gathering medical information. Rather than having a patient first be seen by a clerk — who is interested in details of insurance and other non-medical information — a triage individual should garner the details of the patient's medical history. This reassures the patient that the facility is a medical one rather than a primarily money-making one. It also defuses much patient hostility and improves hospital-community relations.

A discrete resuscitation area is necessary in an emergency department in which more than one resuscitation per day is performed. Such an area should have at least two patient carts. This is especially true in hospitals which see a great deal of automobile trauma, wherein several seriously-injured patients can be expected simultaneously. A room 14 × 17 feet in size is the bare minimum for a resuscitation area. One 18 × 20 feet would be even better, especially in an institution in which a trauma team approach is used and four or five medical people can be expected to be working on one patient at a time. X-ray capabilities can be built into such a room, but this can be quite costly. The walls, ceiling and floor would have to be lead-lined and this would be a major expense. An x-ray unit in immediate proximity to the resuscitation room is less expensive and probably nearly as convenient. I recently visited a university hospital which had constructed an emergency department with four discrete trauma resuscitation rooms, with one bed in each and no communication between the rooms. Such a design is wasteful in terms of the personnel required to manage these rooms. If nothing more, connecting doorways should have been cut to connect the individual trauma rooms. There is no need to establish separate rooms for cardiac and trauma resuscitation. These can be carried out in the same room.

The specialty rooms that many physicians feel are necessary in an emergency department are an eye, ear, nose and throat room, a cast room, a psychiatry room, a social service room, a consultation room, a police and ambulance area, and a decontamination room. There are those who argue that a separate eye, ear, nose and throat room is not essential, since a slit lamp and other equipment can be mounted on wheels and moved from room to room. For best utilization, however, it is best to provide a separate area serving this function. For example, it is necessary to turn off lights for an eye examination and is an inconvenience to darken the entire emergency department to do this. An orthopedic or cast area is an almost universal component of emergency departments. Many physicians argue that a separate psychiatric room is not a necessity. It is true that there need not be a sophisticated facility, but it is very useful to have a locked room in which one can put the violent patient to protect him and prevent him from disturbing the entire area. The need for a social service room is dictated by a hospital's patient population. An institution serving a large number of indigent patients and patients in the geriatric age group will find it convenient to provide an area in which social service can function for these patients' benefit.

A consultation room can be utilized for a number of functions: Physicians can confer with patients' families, chaplains can talk to patients or families, and other interactions can take place. Such a room should be located closer to the waiting room than the treatment area, so that a family does not become entangled in the treatment area. A room for police and ambulance personnel is very important. The police do a good deal of work in the emergency department. They have a tremendous amount of forms to fill out, and it is useful for them to be provided with a separate area in which to do this. It is good to keep police out of the treatment area. They don't belong there and their presence may engender a certain amount of hostility. Now that many emergency departments are involved in the training of paramedics and some even have ambulance personnel stationed in the area, there should be a room where these people may stay and perform their necessary paper work.

A decontamination area should be considered for a planned emergency facility. This is for patients who come in from civil disturbances having been tear-gassed, from industrial accidents having been involved with noxious chemicals, or having suffered nuclear isotopic contamination at a power plant or laboratory. Such patients need to be decontaminated prior to their entry into the emergency department and the rest of the hospital, and a separate area should be dedicated to this function. A major problem with specialty areas is that all such areas in which patients are to be kept require nursing care. If these areas are not convenient to the main nursing station, this will not be well supplied.

The observation-treatment area has recently received some attention. A four to five bed area is probably adequate for most hospitals. The decision to establish such an area depends upon individual needs. All beds should have monitoring capabilities. There should also be toilet facilities and provisions to feed patients in this area. I refer to this as an "observation-treatment" rather than just an "observation" area, as I believe a certain amount of treatment should be performed there as well. Certain treatment is best administered apart from the main emergency area. Asthmatics, for example, are better treated in a quiet room rather than the bustle of a busy emergency room. The essential factor in maintaining such an area is a nursing staff capable of being present there 24 hours a day. The need for physicians to attend to patients in this area must also be taken into account. Many such areas have been established in hospitals with house staff members who can be assigned to observation-treatment areas as part of their duties. Physician staffing in a hospital without such a house staff would be a problem.

Even the smallest emergency department needs to have an x-ray facility as part of it. If the radiology department directly abuts the area and an x-ray room is dedicated to emergency patients, such an arrangement will probably suffice. It is much better to have a facility within the emergency department itself, though, one with two x-ray tables rather than just one.

Some large departments maintain their own laboratory, utilizing automated equipment which requires little space. It is not essential to have such facilities in the department. Their presence raises a number of issues of personnel control, quality control, and administration. Every emergency department should, however, provide a small laboratory for physician use. Such a laboratory should contain a centrifuge, microscope and equipment for performing urinalyses, Gram stains, hematocrits and other similar procedures. All emergency departments need a director's office. There are too many hospitals in which physicians have no place to sit down or perform their necessary paper work. In smaller emergency rooms the physician on call room and the director's office may be combined. Often there is a similar deficiency of administrative space for the head nurse. Such an area must be planned into any new emergency department. In the teaching hospital, each member of the staff who has working and teaching responsibilities should be provided with office space. Such space should be situated within the department rather than elsewhere in the hospital or in another building. Classroom facilities are likewise essential. Such facilities need not be elaborate. A classroom capable of holding thirty people is sufficient for almost any institution. Office space should also be provided for a business manager, assistant administrator, and clerical personnel for the area.

There is not a great deal written about the calculation of emergency department square footage needs. The American College of Emergency Physicians has issued an emergency department management guide in which a calculation is made of a need for one patient bed for each two thousand patient visits per year. This is all right as far as it goes, but it does not translate into square footage. It does not indicate how much space needs to be allocated to x-ray, laboratory or administrative areas. Examination room size in the Management Guide is listed as 10 × 14 feet. This in excess of what is necessary. An 8 × 9 foot room is generally sufficient for rooms other than the resuscitation area. R.B. Rutherford calculated square footage needs at 10,000 square feet per 100 patients treated per day.[1] In this he included administrative areas, x-ray areas, laboratory space, and all other needs. This takes into account the traditional emergency department format, which is designed along multiple specialty areas with many individual cubicles. Such a design wastes a great deal of space. Departments utilizing the central nursing station concept need not construct as many walls and perform well at 6,000 square feet per 100 patients per day. An x-ray facility with two tables and a dark room requires 600-800 square feet. There is a process called daylight processing for x-rays which eliminates the need for a darkroom which might be utilized more in the future. This will result in a saving of space. When volume rises above 100 patient visits per day, thought should be given to a division of patients between the emergency department and an ambulatory care facility. The latter should concern itself with non-emergent medical problems.

Before any rebuilding or major alteration in an emergency department is proposed, the hospital's philosophy should be thoroughly explored. To a great extent, a hospital's administration and medical staff determine the size and type of emergency department that is operated. There are a number of factors that influence staff and administration attitudes toward the emergency department. One is the percent bed occupancy. The hospital which runs an 85-90% bed occupancy on mostly elective admissions will probably be unwilling to build a large department capable of treating a large patient volume. The age of the medical staff is also a factor. If the average hospital staff member is a relatively old physician with an established practice, he will not favor a busy emergency department. He will not want to open his practice to a large number of new patients, and he will not look kindly upon being called to take care of emergency department patients at night. Conversely, if the medical staff is young and the physicians have newly-expanding practices, or the hospital has a low bed occupancy, a large volume of emergency patients to fill such needs will be looked upon favorably.

At teaching hospitals and training institutions, a clinical service may determine the number and type of cases that are appropriate for its training program. If a clinical service has an affinity for certain cases and the emergency department does not supply these — or causes hospital beds to be full so that the services' preferred patients cannot gain hospital admission — that clinical service will actively resist any emergency department expansion.

Finally, the principle of infinite capacity and supersaturation should be taken into account. No hospital, medical staff or emergency department has infinite capacity. They cannot infinitely expand to meet the needs of their population. The emergency physician must come to a realistic appraisal of what his hospital is willing to do and what the medical staff is willing to support. At some point the emergency department will become supersaturated. More patients will apply for treatment than can be treated expeditiously. At this point, waiting time will build and patients will become sufficiently dissatisfied to choose alternate means of medical care. This is primarily the case with the ambulatory type patient who has a non-emergent medical problem. The number of critically ill patients, however, continues to appear at a constant rate. Hospitals that relocate and are unsure of the demands that will be placed upon their emergency department can utilize demographic surveys of their service area and statistics to predict the frequency and type of illness in the area of relocation as the basis of determining the size of the emergency department to be constructed.

Lawrence R. Drury, M.D.

FOOTNOTE

1   Rutherford R.B.: Organization-design, function and operation of outpatient clinics and emergency rooms. Hill G.J. II, (editor): Outpatient Surgery, Philadelphia, W.B. Sanders Co. 1973. pp 4-32.

CHAPTER 9

TRIAGE

Peter Rosen, M.D.

"Triage" is a term derived from the French meaning to sort into categories. It's medical origins are in battlefield medicine, where a distinction is made between soldiers who are dead or inevitably dying, life-threatened but salvageable with care, and minimally injured and readily restorable to combat. In a combat situation primary attention is paid to the third category, because a major war-time goal is returning wounded soldiers to combat as rapidly as possible. Triage has a different orientation in civilian practice. In the specialty-oriented Emergency Department, triage was necessary to direct patients to the appropriate area for treatment. Another form of triage is the separation of patients along other than traditional specialty lines — namely, according to medical needs. This is a more legitimate function of triage.

Such triage serves a reasonable public relations purpose as well. When we designed our triage system at the University of Chicago it was to some extent with this in mind. Under the previous system, the first person who encountered the patient was non-medical. This concerned me, because a clerk does not see a pa-

tient in terms of disease, but in terms of certain
statistics and information: address, insurance
number, and so on. It is the job of the clerk to gather
this information. We decided that we didn't want
clerks triaging our patients because even if they had
been working in the department for many years and
were expert at it, they weren't medical personnel and
had no formal training in the field. While every patient has a right to present for treatment in the
Emergency Department, the physician has the
responsibility of assessing the priority of care, and
thereby determining who should be seen first. This is
important even in an area with a patient load small
enough that patients can be seen sequentially as they
appear. There are a number of Emergency Departments in Chicago that are sufficiently small to allow a
single doctor to easily keep pace with the flow of patients. Even if there is only one patient per hour to be
seen, however, the urgency of the response to that patient must be determined. This is the primary responsibility of triage. The distinction to be made is this:
Does a particular patient have 1) a life-threatening
problem, or 2) is he ill and in need of prompt attention,
or 3) has he a medical problem whose treatment does
not require an immediate response?

At our institution a triage decision placing a patient in the third category would determine that the
patient was seen in an ambulatory clinic geographically contigous with the Emergency Department rather
than in the emergency area proper. Once a chief complaint and a set of vital signs have been acquired by
the triage officer, it is possible to make a decision
regarding where the patient should go and how rapidly his care need be achieved. Another function we
grant to our triage people is the definition of patients
who can be sent to those specialty clinics willing to see
Emergency Department patients without further

screening. In many Emergency Departments, triage persons may institute appropriate diagnostic tests and this may speed patient care.

Who should perform triage? We initially charged physicians with performing this task but this worked very poorly. First of all, physicians are too rare a commodity to spend on triage. Second, physicians will not perform this job for very long without becoming bored. Physicians also overestimate their ability to determine the severity of an illness at a glance. They make far too many errors in this regard. We quickly abandoned physician triage and instituted nurse triage. This was a little better but not a great deal. In our case this was because we were short of nursing personnel and couldn't spare nurses to perform the triage function. Also, nurses were not as consistent as had been hoped in gathering a chief complaint and vital signs because of their desire to get involved in patient care. They, too, felt — erroneously — that they had the expertise to evaluate patients just at a glance. Furthermore, many nurses were afraid to work in our waiting area. Since many of our patients are drunk or hostile, this was not without some justification.

Since there was at this time a large labor pool of ex-military corpsmen available, we decided to train a group of these as triage officers. The reason that these individuals were chosen was that they already possessed a certain amount of medical training and in addition, they had no pre-existing notion of the extent of their jobs. We could therefore, determine what this would be according to our needs, rather than the standing rules of departments, unions, or supervisors. There was an interesting response to the decision to introduce this group as our triage officers. The nurses, who had hitherto resisted all efforts to induce them to perform triage, suddenly became very fond of the job

and couldn't understand why they were no longer asked to do it. This altitude lasted for some months, but
eventually disappeared.

The number of triage personnel required varies
with the number of patient visits. Most Emergency
Departments have peaks of patient flow at
10:00-12:00 in the morning and from 3:00 in the after-
noon until 11:00 at night. Staffing should be designed
to meet the needs of these heavy patient visit times.
At the time we instituted our triage system, our patient load was about 200 patients per 24 hours. We
found that at our peak of activity we were required to
register as many as 20 patients per hour. This caused
a backlog of patients waiting to be triaged. During
these hours we determined that we needed at least two
triage officers and would probably benefit from the
presence of three. On the night shift we needed only
one. This meant a full-time group of at least six triage
officers — and preferably eight.

We have also utilized medical students for this
task. They are usually very enthusiastic and do a
superb job once trained. We have been able to attract
freshman and sophomore medical students predom-
inantly. Since these students have very limited patient contact elsewhere during these years, they love
the job. The only problem with this group is that it is
difficult to get students to keep the medical history
sufficiently brief. Although you would expect a
medical student to be able to take a pulse and blood
pressure, this was not the case and we had to train
them to do so. We had the triage officers themselves
train the students in this. It is a good experience for
medical students to be taught by a nonphysician.

One of the problems our triage officer group encountered was a lack of job mobility. There were no
supervisory positions that they could move up to

which were comparable to nursing supervisory status. In addition, the position of triage officer was unique at our institution and none of the surrounding hospitals had comparable jobs for persons with only military training. Now that combat-experienced military personnel are no longer returning to civilian life, it will be more difficult to acquire appropriately trained personnel to perform triage. It will be difficult to train triage officers without having to develop a time-consuming and demanding educational structure. A logical answer is to utilize paramedics. In most systems paramedics do not work consistent shifts and often take a second job. Triage would be a very appropriate second job for this group.

A great many hospitals have continued to use nurses to perform triage, but that is not a good nursing function. Nurses tend to find triage boring. It is also difficult to control the quality of their triage. One of the benefits of having a designated cadre of people performing triage is that this becomes their primary role in the emergency team. They become very proud of that role and perform it better than any other personnel. Boredom is a problem with this group also. We have attempted to dispel this somewhat by occasionally stationing triage officers in the Emergency Department proper and having them contribute to the general work of the area: taking electrocardiograms, starting intravenous lines, bandaging, and so forth. They have been a great help to us in this capacity and this has served to relieve some of the boredom which develops from long-term commitments to triage exclusively. They must be made to understand that their primary obligation is triage, though, and in order to free them to perform additional functions, additional positions must be available so that the triage area is adequately staffed.

I had originally planned to train the triage officers as physicians' assistants and to use them as patient care personnel caring for non emergency problems in our ambulatory clinic. However, I found the requirements that this would make on our faculty too great to justify and this has not developed.

I have alluded to interpersonal problems which developed between nurses and triage officers. Part of this was due simply to the fact that the triage people represented a new commodity, so this problem subsided after six months to a year. Another problem which developed was one which I call the "Doctor syndrome". The triage officers not only began to think of themselves as doctors, but passed themselves off as such. They are always attired in a white coat and have a stethoscope conspicuously present. While they do refer to themselves as doctors, if a patient should do so they are not quick to correct that mistake. I have no good solutions to this problem except to remind them periodically of what their primary role is and try to reinforce its critical nature. Physician control of triage personnel is essential. There must be a member of the emergency medical staff who is willing to take this as an ongoing responsibility and work with them in a very close supervisory relationship. When I was meeting with the triage group on a regular basis, there were few problems. When I became busy and was able to meet with them only once a quarter problems began to develop.

What is the accuracy of this group's work? We have found that of all the groups performing this job they have made the fewest mistakes. Their level of triage error is about 6%. There have been no disasterous triage mistakes in the 6 years of the program's existence. That is, there has never been a cardiac arrest in our ambulatory clinic. There were errors involving placing patients who were quite ill into the ambulatory clinic because the triage officers did not perceive the degree of the illness. However, there has never been a patient who has died, either on route from the triage desk to another clinic within the hospital or in the ambulatory clinic itself.

Legal considerations are minor. As long as the triage person is operating under the supervision of a physician, he is not practicing medicine. He does not discharge patients from the Emergency Department. Only a physician does this. What he does is determine how rapidly a physician will need to intervene in a patient's medical care.

Since introducing our triage system, we have had better results with our patients regarding their feeling towards the Emergency Department. The first thing that happens to a patient in an Emergency Department seems to be indelibly stamped on his memory. In our department what happens first is that someone asks him what is wrong with him. The first contact the patient has is a medical one rather than a clerical one. We believe that this is essential. We have received many positive statements from patients about our triage people and ours is an area in which few personnel accumulate compliments.